AF888023

F.D. Krill

mind beside itself

Writings at the Edge of Science, Mysticism & Delusion

Wasteland Press
Shelbyville, KY USA
www.wastelandpress.net

Mind Beside Itself
Writings at the Edge of Science, Mysticism & Delusion
by F.D. Krill

Copyright © 2005 F.D. Krill
ALL RIGHTS RESERVED

First Printing – June 2005
ISBN: 978-1-933265-51-3

THIS BOOK IS OPEN IN PART OR WHOLE TO BE REPRODUCED OR DISTRIBUTED FOR ANY NON-COMMERCIAL USE WITHOUT PERMISSION OF THE AUTHOR.

Printed in the U.S.A.

Our reality is as big as our ability to allow for it. To accept any limited view of the universe or God, as the whole is comforting in its security but suffocating to the spirit. It is the equivalent of squeezing our mind into a jar and letting it pickle in its putrefying projections and insistent protections of stale thoughts and rigid actions. Western mind encapsulated at an early age by a trickster god, seals its air holes with runaway rationalism and destructive diversions. Its not that our civilization refuses to offer us important truths to ponder, but it sells these truths as absolutes, dogmatic jars to limit our reach and blur our visions. We accept our managed myopia as shelter from the demons and dragons that prowl the depths, always reluctant to ask too many questions, always scared to look too deeply at the freak-monkey that stares back at us from behind our slanted mirrors, anxious that we haven't bolted all our doors and windows we shudder to think that we might have locked the demon in with us.

This is a lie. In its inability to convey the whole truth it is spreading falsehood. It is in what hasn't been said that reduces this work to mere propaganda. It is a product of an unstable mind born of a broken and extinct culture. See through its asymmetry, for I am unable. It is in what has been forgotten that the seeds of a system's destruction have been implanted. What dark granules of fear sit shadowed by structures, slowly sublimating into these sentences. Question the one that pretends to know, because it is in the unspoken passages that his lies and hypocrisies are most revealed. That too is a lie, for in what is clearly stated and believed exists a delusion of knowledge that is equivalent to a deep ignorance.

Truths, well known to our ancestors are reemerging. The mystical, subjective wholeness of eastern beliefs are reborn in the disguise of quantum mechanics. Flashes of the aboriginal dream-time emerge as the confines of tunnel vision open up to the magic of the periphery. Infinite truths exist for those who are open to them. Our consciousness is a light that shines in the direction it is pointed. Narrow beliefs dim the beam to a world of detached parts. A wider shine reveals the connecting patterns.

The world-views of every culture reveal varying aspects of reality. A balanced knowledge of truth indicates a balanced society. A good way to gauge the accuracy of a paradigm is to know how long a culture survived with those beliefs. Many tribes have survived thousands of years because they developed a great understanding of their living land and have integrated their culture into its cycles. Our society has developed views of reality riddled with half truths and narrow beliefs, resulting in a runaway system out of touch with what we are, what we are within and what's within us.

The clown prophet of Mount Essera He set his craft on sea to know where the current would take him. A jagged trip ravaged his memory, salt spray cleansed the film from his eyes. Here ashore berthed between beached giants he seen the fissure. He traced the fault line to a burgeoning hamlet. There in the earths' soft spot new spires straddled the deep and frightening chasm. Elaborate bridges of rubber stitched sutures and cardiac myofibers held the dream together. A travesty of a river, hardened concrete frozen in poses of anguish had stopped just before reaching the oceans power of redemption. He walked the main thorough fares, fearing his alienness would make of him a spectacle. To his relief and fright the citizens eyes traveled on past him, forever lost in meaningless loops and deductions. Sometimes he would catch a look, curious eyes changing to loathing before quickly turning away. Within his thoughts a debate was instigated. On one side sat Angel Doomsayer facing the enduring smile of Unfounded Optimism. Their voices were heard and acknowledged in the echoey inner soundstage. Since his landing eons and moments before the voices would rage. A mass of confusion and tireless tirades engulfed his will. Little made sense in this pop-up box town with its busy sleepwalkers and their servants. Momentarily blinding himself to the street scene before him he listened in to this debate, he believed it important for it touched on a memory of when he was free from the coming battle. The question addressed the following topic: is the Overmind diseased, or is its suffering a logical reaction to this place and time? The debate, like all others dissolved in the discordant racket of a millennium of hardened experts.

As the dispute drifted to white noise he looked outwards towards a factory. In time he would know this place well as an object of curiosity and indirect income. Here was the center that readied the concrete, which filled the cities divide. Beneath roamed the pulverizers sent out to mine the stone that kept the people occupied. Thousands of lives wore umbilical spokes to the factory's' hub. Only his love of absurdity buffered the beating he took upon entering the mines. Tracing subterranean cave-ins and prophetic tremors to the grand chasm he seen the grounds beneath the city undercut and eaten away. He grabbed at the next person he saw and yelled, "CAN"T YOU SEE?" The strangers' eyes were familiar to another time and lived both lives equally. "Can't you see? Your world is broken in reply to your attempt to fix it". "Well yes", he answered as he turned and walked away.

The ancient principle of chaos is difficult to explain but easy to know. It is too often identified with disorder, something to be controlled and prevented, a dragon in need of slaying. Many have known that behind our contrived man made order lies a much more free and intricately beautiful reality, an infinite well of fantastic imagery flowing and creating rich patterns of liquid music. Surrounding us and within us is this creative flow. It is the brainstorm of swirling ideas, the complex and curious patterns of clouds, the graceful pattern dance of burning smoke, the rolling swirls in our coffee, the movement of celestial spheres. It is unrestricted and undirected thought. Chaos is the well- spring of all creation. It produces the seed patterns from which Gaia grows. Nature is the living condensation of chaos, the forms and patterns chosen then immortalized from the fleeting chaotic fugue. The unfurling fern, the ocelot's spots and the snowflake's shape, share their origin in this cosmic swirl. Mysterious, irrational and complex, chaos embeds reality with layer upon layer of super imposed patterns that intricately unify our universe in every possible dimension. Hidden patterns spread everywhere for those who are open to them, from a gambler's streaks to synchronicity. The infinite fractal complexity of chaos weaves its artistry through the fabric of existence. The identification of these patterns is the motivation behind mathematics and science, as well as art, music, mythology and astrology.

The absence of perfection in nature isn't a sign of corruption. Imperfection allows new movement and growth, preventing monotony. Mechanistic precision and mass production have given our culture the supernatural ability to create forms very close to perfection. Instead of creating heaven on earth, we mold a tupperware hell. Soulless artifacts roll off the assembly line, each the clone of their ideal. Their use shields users from chaos and enforces the idea of objects and objectivity. Given enough time, nature soon softens the edges, rusts an immaculate finish, and crackles designed uniformity. This new- found depth and character of our used belongings often creates social pressure to replace them with the new and shiny. The avoidance of chaos is taught to our young, pushing fears of old forests, flames, deep water, sex and drugs, assuring an unhealthy relationship with the unknown. An overdose of machine perfection and lack of chaos is a recipe for meaninglessness and boredom.

Most natural forms don't conform to the sane and predictable. They emerge far from the deterministic, Euclidian and linear regions of mathematical space we've become secure with. This is the habitat of infinity, where knowledge is always incomplete and everything is possible (or probable). Here lurks what are called chaotic attractors, spirits hiding in the jungle of overlaying patterns. Partially ordered but always wild, these forms are evolving inventions, always different and relative to changes in space and time. Attractors are parallel to archetypes of the collective unconscious, they are the living symbols that underpin reality. Always ready to emerge into conscious light they resonate with the forms that habit has already channeled into existence. They are the basins in which the direction of mind is drawn, the living mold which matter fills, the force behind a wind-sculpted boulder realizing itself, and the negative space of a monkey's ear. An archetype is both the harmony that a pianist seeks and the idea behind the piano's structure.

Deconstruction, far from being just a post-modern, trendy belief system of our day, is the dominant theme of our civilization. Fragmenting and desacralizing reality into digestible bite size pieces. Divinity, deconstructed to the whims of a vengeful bearded-man god, no longer infuses our earth with its holy breath. Nature, when deconstructed to mindless random processes, is cleansed of spirit and meaning and ready to be clear-cut with clear conscience. Mind, when deconstructed and transformed into an illusory construct of a computer brain loses its vital connection to the universe. Dreams, stars, bears, wind, emotions, visions, ancestors, trees, blood, rocks, music, beauty, tears –all that once left us filled with awe, meaning and life has been diminished into some pale shadow of a skeleton of what they once were. A child stares in wonder at a rainbow then ceases to wonder when told that it's just a refraction of the light into different wavelengths while passing through water droplets. The child satisfied with this explanation is a little poorer, a little more broken and a lot more alone. A masterpiece of beauty, a triumph of creation, a holy covenant, a magical alchemy of sun and rain has been killed by logical conclusion, one which never bothers to explain its transcendent beauty and never asks why it should exist.

In the end deconstruction devours itself, and that is its saving grace. Science and religion, perpetrators of the Great Forgetting now find themselves victims of the same process that first empowered them. Waves of meaningless and absurdity are dissolving the systems of control that arose in the absence of knowledge of who we are. Our greatest fears explode into consciousness after eons in the darkness. INSANITY ANARCHY CHAOS APOCOLYPSE ... have we been taught to fear our own savior? The only difference between a Zen master and a psychotic is in the reaction to their breakdown. The Zen master welcomes the disintegration of the imposed order as a liberation from illusion, while the psychotic trembles in fear of his immanent collapse. As the veils are lifted, will we cower in terror at the thought of a vicious judge ordering our damnation, will we turn away in despair at the idea of eternal nothingness or will we acknowledge that those are fears imposed on us by our captors? Awaiting our messiah in a new doctrine or ideology has led humanity further and further into confinement, as the words themselves have formed into the bars of our enclosure. As the meaning of those words dissipate into nonsense the wilderness will be revealed as our guide and mirror, a state of being where there is a order without control and divinity without rule.

Unity is the basis of morality in nature. A being in line with the universal flow is in a state of grace, wholeness or health. Natural order forever moves toward this ideal state in a futile attempt at perfection. It is a dreamers' attempt to fill infinity with ever changing creation, a saturation of possibilities, a mindscape of endless diversity that breathes as one. This seems the ambition of the universe. Pathways of thought and evolution seek a balance between the need for harmony and the momentum towards change. While changes spark creativity, they also stir imbalance until they are further integrated. Anticipating and assimilating changes to prevent imbalance, are functions of learning, memory, intuition and wisdom. With the ceaseless movement of evolutionary flow, it is miraculous that any living being can exist. A cell, a body, a mind, an earth, billions of fragments in constant flux, yet somehow not dissolving into disorder, until the imbalance is too great and the sacred field that imagined this harmony then departs.

As unity dissolves and the field fragments, the soul is weakened. The smaller wholes nestled within the larger grow without limits, no longer integrated into the body that nourishes and sculpts them. When the growth of an individual is out of sync with the aesthetic and survival needs of the whole it becomes diseased. Producing fragmented, insane cells and malignant individuals competing for energy with the body that they reside in.

This process of creating imbalance is iconic of the age that we are living in. Sprawl blights the earth's continents, noxious runoff sickens the sea, smoke stacks belch foul vapors for all to inhale, introduced animals decimate island paradises. The earth's organs are being dismantled and decayed in a million profitable ways. Populations exponentially explode as burgeoning colonies outwardly progress, sinking their power lines into the planet's skin, sucking out juices like an insatiable, mechanized tick. As waters dry up, deserts expand and top soil washes away, we the plunderers, having shit in our own nest search for more balanced frontiers to suck dry.

Imbalances in our surroundings are reflected in our bodies. Where the land is decayed and people are marginalized, the unity of bodies begins to weaken, as out of control organisms grow unchecked from within. Infectious boils, blight, rot, crust – ambassadors of imbalance attack alarmed organs. In places where stress abounds and synthetic chemicals enter nature's cycles, the body attacks itself. Immune systems facing unknown and unprecedented adversaries sink teeth into their own flesh. A paranoid watch dog in a house of invisible thieves. Arthritis, lupus, allergies, migraines, chronic fatigue, MS, Alzheimer's, Lymes, AIDS, and an endless number of syndromes of debilitating incurable disease of disharmony abound. Cancer, the archetypical symbol of a body at war with itself, symbolizes out of control growth, the logical outcome of the never- ending conquest of land, energy and power. The analogy between human empire and malignant tumors is too powerful to resist, an analogy often avoided in polite and civilized conversation. It comes across as hateful misanthropy. At first thought it seems to label us humans as ruthless evildoers, but this is not inevitably so.

A cancer cell is really no different than other cells of the body. Its similarity to other cells is what makes it so hard to destroy. The difference is not in what it is, or even what it does, but in its lack of coherence with the body. It exists out of step with the world around it. It creates its own disconnected world with an intelligence of its own. As it spreads like a brush fire, it reroutes blood supplies, to help feed its limitless appetite, forms membranes to protect itself from integration or attack and creates compounds that deny its mortality. A cancer cells' destructiveness isn't due to the existence of any of its qualities. Such things as growth, greed, individuality, competition, and selfishness are also characteristics of healthy beings. Its' exaggerated expression of its qualities is the problem. It is out of sync and therefore at odds with its environment.

The timing of a cancer cells' actions are off. It exists in a fallen, dissociated state. Its alienation from the body's cycles and rhythms can only lead to the destruction of its world, and therefore itself.

Science often has difficulty understanding cancer because it has a warped view of nature. Seen from a neo-Darwinist perspective everything should be a cancer. In a thinly veiled rationalization and projection of our destructiveness, we are told that self interest rules, that our earth is composed of billions of fragmented parts battling with each other to pass on their genes, a Hobbesian war of all against all. We spread the word that Mother Earth is actually a dead rock, no coherence, compassion or altruism. *Hey! Its survival of the fittest, and we humans won. Its nature's way, we're just robots anyway, we're programmed for conquest. There is no right or wrong, it's a meaningless world, what the fuck does it matter anyway?* Religions rationalize our conquest as God's will. *We must tame and order the world, have dominion, be fruitful and multiply. Our world and our bodies are evil, to hell with the Earth, we're going to heaven afterwards.*

What the hell are we doing? What is wrong with us? Have we been blinded? Bought and sold? Domesticated? Controlled? We are destroying our world. The idea, though quite obvious, fails to induce a response. Our minds brush out the statement with the indifference of a microchip and the rationality of murder-suicide. We see that our current path is a literal dead-end, an instant flash of hell dissolving our world at best or the slow burn of corrosive fumes and unquenched thirst of an unlivable planet at worst.

It's not too late to change paths, but something blocks our ability to turn. Hope for our future sparkles at the corner of our vision, but we are unable to turn our heads. Our death march quickens, while our minds are burdened with meaningless busywork. Fooled, indoctrinated, distracted, stylized, vaccinated, irradiated, labeled, corrupted, encapsulated, domesticated, herded, diverted, perverted, reordered, we keep up with the progression of our demise. We are destroying our world. I wish I could tell you why. I cling on to hope, an elusive entity, a tender tether holding things together. I'm speaking, now to you, my pen funneling and distilling a cry, not just for me but for you, our kind, our being, our Earth. Ramblings flow from a mixture of rage and hope from a lonely being shipwrecked on a self-destructive planet with a self-deceiving mind, an eyeball in a pressure cooker poised to burst.

Governments, corporation, and religions repeat their mantra: *grow, grow grow*, at any expense. Keep growing, for that which doesn't grow, soon dies. Mind control experts of Madison Avenue, remind us that too much isn't enough – more shoes, bigger bank accounts, roomier houses and richer cakes. If something is good, more of it is always better. This is the linear philosophy of abstract concepts such as money, comfort, health and religion. It is the world-view of decadence, it is the battle cry of cancer. It is at odds with the wisdom of nature and of the body – that if something is good, too much of it will eventually be toxic. As we dig deeper to find the roots of thought forms that have materialized into nuclear war-heads, pesticides and strip mines, we find an imbalance of mind. Belief systems that are out of touch with the mind of the earth and universe proliferate, as revelation of truth comes from books, TV, computers, and other man made sources instead of from the natural world. As the mind of humanity becomes increasingly enclosed and self-referential it stews in a stagnant broth of half-truths and fears. A fragile cord that is the conscience of man-kind links us to the greater whole. Artists, mystics, meditators, naturalists, natives, and freethinkers, play the part of the shaman to ailing humanity, providing a crucial link with the universal flow. Those that seek inner balance, help to balance the outer. Most people however, only steady themselves in accordance with our insane society. Playing by the rules, doing what we are told, and seeking the American dream, we become inadvertent destroyers, spreading poison and making the big bucks, while congratulating ourselves as to how far we've progressed. None of us are immune. As long as humanity is dissociated, individual members will also be. The split is reflected not only in our bodies, but also in our minds. The battle lines of man versus nature are drawn in the spaces between our thoughts. Insanity is cancer of mind. Fragmenting our world as we chop away at ourselves, we no longer know who we are, or how to live. We've become lost and nearly every route that we are told to travel has become corrupted, as cancer too has its own immune system. Those that still have a mind of their own, seek to find themselves. Those that are wiser, seek it outside of society's suffocating membrane.

We can seek what is pure, what is healing, what can make us whole again. It isn't some prepackaged, mass produced product we can find in the supermarket or bookstore. It flows from within us. A divine pathway that is unique to every being yet is one with itself. The beautiful result of this flow is evident anyplace man has relinquished control. Untamed wilderness spontaneously moves towards healing. This we can find from within. We can abandon the pathways that end in fences, dams and cages, and let them grow wild. We can free ourselves in the savannas and untamed rivers inside, and watch as things miraculously work themselves out. Healing and wholeness are achieved not through order and control, but by the lack of it. Rationally this doesn't register, but yet a forests leaves bud, the water and air filter themselves clean and a buffalo calf grows up without our help and deliberation. There is tremendous optimism in this as all things eventually move towards harmony. Our bodies perform intricately and flawlessly without our conscious monitoring and control. We can't really explain why its so, but its true. Our minds, if only we allowed for it function in the same way.

I wish to bear witness to the events that have occurred, a play-by-play of the hijacking of a planet. But do you remember how my own memories have been tainted by lies and exaggeration, so too the memoirs of Man have been veiled. Engineered might be the better term. A consciously manipulated mythogenisis streaming into wired minds. Somewhere along the line, the vacuum tubes and plasma screens we stared at became more real then the flesh, from then on who is to say what once existed. Hack writers rewrote our memories and programmed our dreams. They erased our souls and replaced them with pixilated facsimiles. Our blistered outer and inner worlds echo each other, egalitarian Earthly nature dethroned and replaced by a mad ego. History lassoed by the Universal antagonist and made a servant in his court.

Once as part of a great youth revolt we tried breaking out. We swore allegiance to our dying planet, we fought soulless robots in the shadows and we kicked in our glowing screens. We were infiltrated from the beginning, the screens we destroyed in our outer life glowed bright in our inner. Without Them telling us who we were, we no longer knew. Without an identity and with no place left to hide, we couldn't hold out. We blindly ripped at the wires that then grew back even faster. We hated the leaders that brought us here, only to find that they were actors, cast in these roles for us to despise. Whatever we attempted to think up, someone had been there and trademarked the idea. The pathways of thought were channeled into rooms corrupted with the filth of falsified images and engineered idols, prepackaged rebellion; one more product for us to consume.

Perhaps all is lost, but I swear I can hear the system cracking under the weight of its own success. If the 10,000 year story we are in is one of slavery and emancipation, the climax is coming fast, and the collapse is immanent. When it falls it's gonna fall hard. It's a civilization of rot that feeds on imbalance and these are the years of gluttony.

I could never pin down any one way to be. Every face that I've tried on is fleeting and sprinkled with irony. Masks are in. We're all asked to carve a few. One to take to work, one when you're out with friends and one to leave at home. The mighty and Fabled Korporate Kingdom rewards those that can carve the mask most solid. The greatest Game players get the key positions. Those that try to look behind the masks crack the one they're wearing, making them unpredictable and defiant. They are soon targeted for elimination. For most, faces are stamped on in various stations of Mass Human Production assembly lines, called schools and offices. When every pigeon is holed, contestants grudgingly accept their consolation prize; a brand new entertainment center\ alter to serve the gods. A little world handed to you in a box. Why fret about your dreams drying up, when these nice broadcasting folks have made for us some spiffy replacements. Invisible ray-guns fired by geo-synchronous satellites emit coded light rays through thousand mile glass filaments, photons beamed in precise patterns produce reactions in the retina, igniting electric pulses through optic nerves, releasing sequences of neuron transmitters between waiting dendrites, finally neuronal connections implant corporate board-room created gods into our inner mindscape. An army of experts enter our being, speaking their practical inanities whenever we attempt to make a decision. A pantheon of deities leak into our life-field and demand attention and worship. Unlike past gods these aren't as much projections of our soul onto the sky, but the sky projecting gods into our soul in order sell us stuff. Best of all the new, improved and implanted gods are all actors, just like us they wear masks. It's great to be fake, no more pressure to be your Self or need to discover who you are or why you are here. Just pick a set of Role Models and persistently practice impersonating their personas. In the World in a Box the façade is the All, and the ego is One. The rock-solid surface we create hides something more frightening then the nothing we all fear, that is the World Consuming Robot. 9 out of 10 nihilists prefer the idea of ultimate reality being a great emptiness rather then the Divine Executive Sales Manager. One wears black and the other tries to suck out your soul.

Growing branches divide and stretch across the sky searching for a place in the light. Meanwhile the roots seek the darkness as they burrow and feel around obstacles seeking water. Wherever the creativity of thought and evolution send their branches of possibilities, the after image of the tree remains to give unity to all of its seemingly separate parts. The heart with all of its arteries and capillaries is one being searching out cells to nourish. Similar growth spreads nerves and lymph vessels across the body. This kind of branching creates its image in space, other kinds leave their impressions in time. The family tree connects beings across time, narrowing relatives towards a common ancestor. The evolution of all earth beings can be traced backwards to ever more general classifications and simper patterns toward a common ancestor. This ancestor and its near infinite descendants exist as a single entity, unifying all life on earth as a single organism. Without the added dimension of time, all we see of our family tree are its terminal branches, each appearing independent of one another. This tree of earth evolution further exists as a branch in even larger unities of thought, giving a wholeness to all of existence. A cross section of any branch reveals the past buried inside the present, like rings on a tree. The sapling lies buried within a giant oak, as the center of each layer of concentric growth surrounds it. The universal sapling contains the earliest formed habits. These became the laws of nature that sustain the outer growth of patterns that form into matter. The stable patterns that form the trunk, create the frame work for all further branching of evolution. The tree and its branches represent all that has already been created – the living past. The present exists as the billions of terminal branches and rootlets that are still extending themselves deeper into space. Limbs that lack vital growth tissue become extinct lines of thought and soon wither away. The ends of growing twigs are covered by lush growth of virtual capillaries extending out in all directions. These extensions are relatively fleeting. A profusion of ideas, they snake out into the quantum ether and disappear until a selected pathway is chosen out of the many. An idea, when realized becomes a bridge towards further creativity. The earth is an extremely active and blossoming branch connected and nourished by the universe. What we call life is the extending limbs of the diversity of our planet. The first cell was the sprouting of an idea that all future earth beings grew out from and are forever part of.

The forests, marshes and seas are the earth's vital organs. The living landscapes are the heart and lungs of the world, focal points of respiration, metabolism, photosynthesis, circulation, filtration, transport and detoxification.

Ponder the ocean's reefs. The reef community pulses with a vast array of life, each creature helping the reef to function as a whole. Fish, coral, plants and microbes interweave their lives into a complex physiology. Reproduction occurs simultaneously for many reef beings, revealing the hidden coherence that patterns individual existence. Each coral polyp exists somewhat independently, yet form complex colonial structures beautiful in their depth and symmetry. These structures grow out from each other creating an even larger body sometimes thousands of miles long. Echoing evolution itself, each layer of growth is formed on the backs of previous generations.

The reef evolves as a single organism and also an organ of the earth body. One of its functions is to keep the chemical and mineral composition of the ocean at equilibrium. Reefs deposit an enormous amount of calcium carbonate and silicates as stone, reducing excess minerals brought to the ocean by rivers. They also serve as a filter, keeping the water clean for the passage of light. Coral polyps contain photosynthetic cyanobacteria in their tissues that create food from sunlight. Reef building, affects the earths' floating crust, perhaps even driving plate movement, mountain building and volcanic events. It is possible that our planet moves its continents by depositing rock in specific locations through the action of coral. Reefs create islands and serve to prevent erosion by waves. Besides physically stopping the waves, they have microorganisms that secrete a layer of lipids to change the surface tension of water to prevent wave damage. One of the most important functions of the reef organ is to form evaporative lagoons. These lagoons deposit salt on the continental margins. Communities of microorganisms prevent the salt from dissolving back into the sea by forming a mat over the crystals. The salinity of the oceans have remained remarkably stable partly due to the actions of tiny corals. Testament to this stability is found in our blood, which carries the same salinity of the seas we've evolved from hundreds of millions of years past.

It is becoming increasingly clear that the Earth body regulates the composition of the atmosphere and oceans. It deposits minerals and gasses in its crust to keep itself at an evolving equilibrium. Carbon is removed from the atmosphere by marsh plants and deposited in the ground to regulate the planet's temperature as the sun's output increases. Salts, calcium and silica are placed in the earths' crust by sea life to balance the oceans' consistency.

Water and mineral cycles have evolved to sustain the earth's future. Oxygen, which allows for such things as breathing, blue skies and fire is regulated by the forests and seas. Levels are kept steady, as too much would make the skies burn and too little would suffocate all that breathe.

It is probable that the earth influences many more properties then is currently being researched. The role that our planet plays in its own evolution may include its rate of rotation, revolution, and inclination. Slowly altering its position in space to create greater diversity of life. It is possible that the earth body can also change its magnetic field, position of its poles, speed of rotation and its distance from the other planets and sun. To what degree that the planet's physiology regulates its geology remains a mystery.

Dangers from outside our world must also be protected against. The fact that our planet isn't heavily pock-marked and cratered testifies to the success of its protective shield of ionizing and oxidative gasses which destroy almost all incoming meteors. Our earth also protects its fragile skin by blocking out deadly radiation. This is achieved by the magnetic field that projects its invisible arms across the heavens, blocking gamma waves arriving from distant, imploding worlds. A layer of ozone further protects by absorbing ultra-violet rays flowing from the sun.

Over eons of evolution our living world has learned to survive many types of deadly catastrophes, including chaotic occurrences such as meteors, comets and local supernovas, earth changes such as volcanoes and pole reversals, and sun changes such as solar flares and sunspots. Wisdom gained from these events has made the earth stronger and better prepared for these actions in the future.

Wisdom immanent in the body is evidence of intelligence and memory in evolution. Immune responses evolve even over the lifetime of a single being. The body learns to recognize hostile forces and creates methods and molecules to integrate or eliminate them. Improvements are passed on to offspring and into the collective mind field.

In many instances seemingly catastrophic events become something positive. The recolonization of land following volcanoes, floods, storms and fires create new and diverse ways of life on replenished ground. These events have repeated themselves countless times in history. The Earth learns to influence and predict these occurrences and integrate them into her cycles.

The damage caused by humans is different, as it appears to be unprecedented in the geologic records. The takeover of the planet by a single species seems to be an event unheard of in billions of years of evolution. The speed of which the coup has occurred is a blink of the eye in geologic time, making it all the more deadly. Furthermore the attack is on all fronts. Every one of the earth's cycles and functions are being controlled or derailed in ways that our planet has never experienced. Without any immunity to these types of assaults it is difficult for the earth to mend. Similarly, our own bodies are unable to deal with all of the synthetics we've put into them, so our immune system breaks down or attacks itself. The dissociation of our minds and bodies is the logical result of the attack on our world, as the earth is the extension of our selves. We've detonated hundreds of nuclear bombs, punched holes in the atmosphere and poisoned the air and water in an all out attack on our larger body. The earth has enormous healing potential, but our runaway evolution prevents our integration.

What can be said of being conceived into existence that is immersed in paradox? For every truth that we may utter, the opposing lie has some validity. Often lies contain more truth then facts, thus giving root to mythology.

Because every rule that we could imagine has its exception, the existence of exceptions becomes the only rule. Perhaps that isn't true either. Maybe nothing written in this book is true, it is a work of fiction. Would that make it any less valid?

It seems to be a world made up of partial truths, which cannot be added up to form the Absolute.

Somebody once claimed that the final act of reason is to know the limitations of itself. Transcending reason leaves the mind untethered and life becomes a subversive act of faith that is not anchored in belief.

The only way for the part to imagine the whole is to allow every aspect of it into our minds in the same instant, and therefore to become it.

When we are narrowed into the near blindness of the particular, we are reduced to living in time. From this perspective we are left with puny models of the universe and entrenched philosophers endlessly arguing tired theories.

The thoughts of what something is, influences then becomes part of its character. Whatever is projected is also absorbed. It is the way that our blessings and curses materialize. The acts of defining and labeling serve to narrow a belief to the values described. A belief about what something is, reflects these ideas back to the believer. You see what you want to see, you are shown what you take to be real. Seeing is not believing, believing is seeing. The more popular a belief the greater is its influence in any particular region or time. The repressed aspects of any belief system exist in the shadows of the mind and eternally seek to balance out half-truths.

Who is more real, a fictional character or a being of bone and flesh. When lies become myths, become gods, wouldn't the imaginary be infinitely more powerful then the actual? If a god becomes the starring character in the life of a man, the man is ruled by the imaginary. But what if the imaginary were real and alive and can only realize itself through the minds of its believers? Would it not create a world of paradoxical possession, where even the non-believers who reject the given gods' powers help in its evolution by making it more elusive.

With an observer created reality we don't just see what we want to see, we are shown what our theories would predict. What a joke on the scientist who believes that his experiments are revealing objective facts rather then a projection of his beliefs and fears.

A question arises. Were Newton and Descartes creators of the reality system so many of us now live under or were they prophets identifying a paradigm that was coming into being? Did the prophets destroy our world with their cynicism or did they merely trace our destructive trends to their logical conclusions.

As our dream of the world unfolds, we may wonder how we fell under the spell cast by the Rationalist who is the god of the Scientist, nemesis to the Poet, divine mechanist to the Deist, straight-man to the Ironist and all around ambassador of obsessive order.

It is a popular trend to deconstruct human thought to meaninglessness. It is to believe that our perceptions of this world are not of this world. Despite this, every subject that we can conceive of, such as music, history or math, are not just aspects of human society, they are mental pathways of Universal Mind. Every person and culture, due to their unique history and spirit reveal aspects of the great mystery that we take part in. An artists' revelation of truth might be through beauty, a mathematician through number. An actress will have revealed to herself, the wonderful drama in life. A gourmet may know God as the ultimate chef. A basketball player may see life as teamwork and competition. A technician might see the solar system as clockwork, whereas a cellist hears the music of the spheres. To accept any one truth as absolute is to worship the part as the whole.

Stones, plants and animals also reveal clues to hidden realities. Shapes, textures, colors and behaviors are filled with rich symbolism of the powers flowing through them. This vital flow manifests itself as mortal forms for fleeting glances then drains back into the mysterious realm that creates and sustains it. Things that we might consider lifeless and unimportant are answers waiting for questions, living evolving bits containing the whole within. If anything that exists is thought to be worthless or irrelevant, it is only a symptom of the times that we live in. Religions have divided the world into the sacred and secular, the divine was believed to stand outside rather then within. Finally it was banished altogether, leaving us as empty shells struggling in a cold and indifferent universe.

Scientists are taught to avoid anthropomorphism and analogy. By neglecting these tools of thought they assume a supernatural premise, that we humans are not of this universe, that mind is unique to us and non-existent in the rest of creation. This dogmatic and somewhat solipsistic assumption leads to senseless theories of a self-assembling world, randomly forming and blindly selecting. These theories create a worldview not in keeping with what we perceive around us. To believe this is to close our eyes to our brilliantly crafted world and to imagine our own thoughts to be an illusion. This runs counter to the wisdom of the ancients; that to know ones' self is to know all. For what is our body but flesh of the galaxies. What is our mind, but an extension of the flow that shapes the universe? Most scientists fail to notice that their ability to predict and to gain knowledge is anthropomorphic. The mathematical patterns that they project on the world exist in their minds as well as on the outside. Some of the greatest minds have been troubled that the universe is at all understandable. Most will avoid wondering about it, afraid of tumbling into the endless sea of mysticism. The physicist Stephen Hawking wrote "theory exists only in our mind, and does not have any other reality". But if this were true then why does it fit. If thought pathways that form mathematics and logic exist outside of our minds, then science must concede that the universe has mental processes synchronous to human thought.

If approached with a gaping mind, nature reveals itself not just as a logical system but also as a charmed mirror, reflecting back to us what we believe and answering any questions that we may ask. By simply accepting the possibility of a sentient world our minds eye inserts ourselves deeper into the entity that forms and nourishes us.

An organism develops by retracing the steps of its ancestors. Beings resonate with the pathway of their particular type. Environmental and genetic influences create variations that then become part of the morphic field of the species. These varieties influence future organisms, which develop along similar pathways and outside influences. The more often a pathway is retraced the more likely it is to be followed during development. The final forms and their developmental pathways are attractors towards which evolution creates its forms. The memory of a life form is a part of the memory of its species. As beings encounter similar life circumstances, habits are retained in the memory of the species to deal with them. These habits are preserved as instincts and rituals. These rituals are celebrations of the cycle of being, remnants of the creative spark that initiate a pathway into existence.

The terms morphic field, attractor, idea or spirit are similar concepts. They represent the living formative principle that channels thought towards the material realization of itself. All things and thoughts are channeled towards these patterns. Seemingly random behavior on one scale is structure and pattern on another. Our universe can be known as a being creating itself in the image of its own thoughts. The laws of nature are the most powerful and consistent of ideas, but they too are evolving concepts. They are the instincts and rituals in which all life tends to abide.

The universe is a vast embryo taking form. Gravity is a vast organizing principle that positions matter in its proper place in the body of space-time. An unnamed force propels matter away from certain regions of space. Matter conglomerates in some areas directed by the field and accelerates away from others. Cosmology and embryology are concepts differing mainly in scale.

Mind is evident in the body's ability to integrate billions of cells. Each cell has its own birth, growth and death. The body learns to adapt to internal stress and external pressure. A weight lifter's muscles expand, bones grow dense and the heart and its vessels become more efficient. A lifeguard evolves dark skin, a villain a scowl, a firewalker calluses and a gymnast flexibility. The body notes every motion and reacts towards recreating itself as best as its wisdom allows. The memory of past events shape and scar the field of the soul to deal with life's situations. This evolving field then works toward shaping the next generation, passing the lessons it learned to offspring who are then predisposed towards the lifestyle of their ancestry. A species evolves through the needs and desires of its members, communities and planet, in a complex web of dialogue within a growing mind. All boundaries are permeable and transient and all parts are also wholes.

A living dream awakens the day. The playful chattering of birds celebrating a spring morning, harsh anger of a thunderstorm, or the melancholy howl of a coyote calling out the night resonate deep within us. The outer and inner touch. We've evolved in a setting, every bit as alive with thought and emotion as that which we now hold from within. We travel in a mindscape in which every turn offers us a connection, a symbol, a bit of truth familiar yet mysterious. The haunting flute of a hermit thrush, an old tree bending and crackling in the wind, a new skinned hatchling snake peering through a fractured shell, a trillion spirits growing and flowing, reflecting and revealing symbols, ideas of an unseeable life-force that animates and lives through them.

Each landscape evolves with its own personality and artistic style. Cultures and species flower out of the ground that nourishes and surrounds them. Further sculpted by wind, rain and stone, creatures form as mental outcroppings of the land. Ideally every growth and movement becomes a beautiful dance, artistically dazzling with a balance of form and color. With diverse expression, spirits of the earth paint saguaro cactuses across the Sonaran, zebras stampeding the Serengeti, polar bears diving in pools of crystalline blue, bright red ape men swinging past strange fruits in what was the forests of Borneo. Big horn sheep stoically gaze across sun set mesas. Salmon wear the water's shimmer as they race to their birth land.

Each environment is a portion of mind, a living stage portraying an artist's vision. This revives the ancient idea of genius loci, the local spirit of place, which helps shape the character and appearance of its landscape and inhabitants. This is the morphic field that draws things towards coherence. It is how an ecosystem works as a single body, consistent in its aesthetics, physiology and logic. The genius loci bestows a sense of place and provides a state of mind to the beings within. An animal, perhaps a rabbit, is a sub mind to the genius loci and also a part of the mind of its species. This greater Rabbit spirit evolves through the experience of its cell rabbits each shaped by its environment. Many tribal peoples were able to communicate and understand these greater spirits: Raven, Bear, River, or Mountain is a representative of its kind. By knowing the land, one can know the inner workings of their own mind. A community of beings interacting and striving towards wholeness loosely defines our bodies, planet and selves. We empower different parts of ourselves as we create our lives. We too are genius loci of inner villages, forests and oceans.

The concept of genius loci accounts for the homing instinct. It is a powerful attractor that draws a being towards home and to a place in its community. Sea turtles swim for thousands of miles to bury their eggs in the beaches of their birth. Similarly living myths draw us towards archetypical stories homing in on classical dramas. Intuition and prophecy are achieved by being able to foretell the plot by recognizing the story lines. Eventually they are immortalized as instinct. Landscapes reenact local myths, revealing powerful symbolism carved into the earth by the chisel of cycles and rhythm. Many of nature's patterns are familiar as symbols used in local ceremonies. Ideally, cultures flow with the landscape as a single entity. The land itself is a character representing a particular state of mind of our planet.

Not only is space unified by guiding attractors but so is time. The zeitgeist is the ruling spirit of a time, shaping land, body and mind according to the motivations and spirit of the era. It unites minds towards a common movement, shaping each generation with the experiences, visions and mannerisms of a living age.

Once in a moment of epiphany among the coral reefs of the red sea I was shown the absurdity of what I was taught in school. How could blind forces of what is called natural selection explain the fantastic forms, colors, patterns, movements and overall coherence of the reef? Far from being drab utilitarian gene replicators, the reef's inhabitants were heavenly visions, spirits of the most magical dream, dressed in their most brilliant finery. This wasn't a world ruled and constructed by deterministic forces, but by creativity, aesthetics, humor and drama. Just as the Sistine Chapel isn't a simple shelter, the reef doesn't exist just to function, but to extravagantly transcend functionalism. Luminescent jellyfish pulse patterns of light across their ghostly body, bright red striped clown fish hide in the protective tentacles of even brighter anemone. Octopus wearing their thoughts on the outside of their body change colors, patterns and form in an orgy of self-expression. The flow of evolution is encouraged towards that which is beautiful.

Evolutionary theory is not new. It was anticipated by the ancient Greeks and is an essential aspect of Taoism, Buddhism, Hinduism, and Confucianism. What makes Darwinian evolution unique is that it lacks a creative principle and is based on chance and necessity. Random events replace design as the formative cause in nature. It is a theory that only makes sense if it's not thought about too much. Ironically, the theory of evolution itself has barely evolved in 150 years as Darwinian doctrines are rarely questioned by biologists.

As a world-view and a religion, Darwinism is nihilistic in character, stripping life of purpose, meaning and beauty as it suffocates the spirit. Wonders of nature soon become market commodities as the world is debased and desacralized. Jaques Monod revealed this lack of meaning resulting from a belief system based on random events, "The ancient covenant is in pieces, man at last knows that he is alone in the unfeeling vastness of the universe, out of which he emerged by chance'.

The existence of chance events certainly doesn't deny purpose and meaning. Thought and evolution are stochastic processes, having a random component, along with an ability to select among choices. Selection also has a random aspect as some things evolve by random arrangement, while other things develop by implicit choice or even providence. Chance is the arrival of the unexpected, it is the wellspring of imagination and play. Its role in evolution parallels its part in creative thought, music and art. Mind and evolution dip their spoon in the serendipitous bowl of chaos and create the New.

Denial of mind in nature obscures the larger picture. The universe is accessible to us because our minds are continuous with it. Our traits cannot occur from a vacuum, if we are intelligent, then what we are a part of must therefore also be.

Mind creates reality, without it time ceases to exist. It is mind that keeps track. We know that it exists everything else is suspect. Mind lends purpose, meaning and justice to all it perceives. In its absence is nothingness. It imparts a motive for existence. For why should we have a universe that no one knows about, and no one appreciates? It explains how things gain in complexity and learn from the past. It is overwhelmingly evident in the extreme precision of natures' laws and in our beautifully crafted bodies. It is not just a ruling mind from above, but also the infinite multiplicity of mind from within. It is both immanent and transcendent, learning and creating, gaining diversity and wisdom as it discovers itself through its works.

Mind has the capability of uniting to create larger wholes. Worlds within worlds, minds within minds each level shaping those above and below. At every center also exists the whole. The earth and the body are analogous, each having highly integrated cells and tissues. Even a cell is an ecosystem of ribosome, mitochondria and chloroplasts, each with their own inner structure, organs and DNA. Evolving and reproducing on their own, yet staying in harmony with that which they are a part of. The earth body, due to its enormous stability and diversity is less integrated than the human body. This allows for greater experimentation and freedom among its constituents. Where the needs of utility are met, free will reigns and the artist is unleashed. Splashes of color, brilliant patterns, divine dances, music, humor and play are signs of a healthy world. Imagination redefines "'survival of the fittest" as the idea of fitness is replaced by that of grace.

Textbooks emphasize the clash between Darwinism and bible believers. It is characterized as an epic battle pitting truth against doctrine and blind faith. Darwinism soon monopolized the concept of evolution and biblical literalism dominated creationism. People chose their preferred dogma, created battle lines and fired their shots. The old war still rages on. This split didn't always exist. In the early 19th century Lamarck proposed his theory of transformism, where immanent mind inherent in nature drives what was later called evolution. Many others speculated on design and imagination as natural process. Even Alfred Wallace, evolutions' forgotten cofounder became a spiritualist later in life. Creationism and evolution, far from being competing theories embrace and demand each other, fitting together like puzzle pieces. One proposes the idea of a creator of a world that doesn't evolve, the other of an evolving world without intelligence. Both require a huge leap of faith along with the suspension of critical thought. Merged together, they bridge the gap between western science and all other belief systems, empirical data and sensory experience, the mundane and the divine.

The years that followed Lamarcks' theory saw the rise of the industrial revolution, setting up the atmosphere for Darwin's Origin of Species (1859). Mind was excluded as an explanation for evolution. Logical and quantitative explanations dominated the scene as the mind of man was immersed in the machine. The popular image created by the science of the time was of a universe where everything was predictable, hierarchal, mechanical and subject to control. God was dying, nature was blind and man would soon rule supreme. A mechanistic view of reality fit neatly with the new ethics of industrialization. Greed, ruthlessness, slavery, colonialism and genocide were rationalized and often encouraged under the new philosophy of social Darwinism. It is the preferred ideology of the robber baron, the corporation and the Nazi. Much of Darwin's text was based on the economic theories of Adam Smith and also had a great influence on Karl Marx. The desouling of the natural world was necessary for both Marxism and capitalism to thrive. What was once a divine creation became nothing but a resource to be plundered. It continues to be taught today unquestioned, a justification for the disemboweling of our world. Evolutionary biology is stuck in a time warp. Textbooks seem like remnants from a nineteenth century reality. Untouched by the mysticism of modern physics, neo-Darwinists still hold to a mechanical and deterministic paradigm of reality. They seek a logical reason for every adaptation and obsessively try to fit their hierarchal models on an essentially egalitarian system.

A story engraved in stone had turned the two great mythologies on their heads. In dense layers of frozen time it is written that the ancestry of all the worlds creatures can be traced to a creative outburst that occurred around 500 million B.C. Encased in rock ancient beings beckoned for acknowledgement. To account for their reality, one would strive to imagine that the tree of life, has a shadow world where the leaves grow the branches and thoughts grow the thinker

Can we still believe in progress while inhabiting a self-destructive world? From many points of view the earth is quickly degrading, as evolutions' clock appears to spin backwards. Hopelessness seems a logical reaction when things aren't improving. The specter of death has always been alleviated through hope in future generations, but what happens when the future existence of our offspring is no longer guaranteed.

The vast rocky desert of mars stares it's hollow gaze back at us through our wandering probes, testament to a once flowing planet laid to waste. In mars we see a warning or a premonition, or at the very least a reflection of our own worlds' fragility. Oceans eternally frozen in rock, a rich atmosphere blown deep into space, a skeletal dryness where natures' artistry once performed. There we see evolution replaced by entropy, birdsong by cratered stillness.

A trick of time, we're caught in the death swing of the pendulum. We view the world either looking backwards towards a golden era or forward to a savior that arrives to late. As sirens wail we place our hope in a technology that has been our undoing. As things improve for a privileged few the rest of the planet foots the bill. Suppressed chaos gushes forth through the cracks in our battered and bloated structure and we call for even greater control to wall the incoming tide. If we release the hold we crash, if we stay the course we only gain an even deadlier momentum.

Progress is the theme of our evolutionary mythology, a linear progression in diversity, complexity and consciousness. But have we forgotten the great extinctions. A witness arriving shortly after the implosion of the age of reptiles, might not believe that time assures advancement. A man losing his family to a terrible sickness may not see that life moves toward healing. Even still springtime arrives and new life stretches toward the light.

We've been told that our civilizations are advanced compared to the primitive savages we had been, we're told that we've progressed and left our undeveloped selves behind. Can we still believe this as we near the endgame of our development.

We search for the missing link between apes and man. Perhaps our obsession with our own progress has made us too blind to see that chimpanzees may have evolved from ancient humans and have returned to the trees. Meanwhile chimps too have been attacked and dispossessed by our progress. As they're fractured cultures follow our lead and make war on each other, what does remain clear is our worlds' penchant for irony.

Darwinian theories have long been used to justify the system of oppression. It is a mythology spun on an evolutionary scale. The central theme presented to us is one of progress. Mankind is placed at the height of evolution, and our ruthless leaders at the very pinnacle of mankind. Evolutionary timelines invariably show the progression of life upwards from the slime. Fish mutate into salamanders into mice into monkeys. At the far end of the journey the hero emerges; civilized man. We are presented to ourselves as the successful fulfillment of life's' story. Each step of the way is presented as a conquest, a thinly veiled rationalization for imperialism and genocide. Amphibians are said to have conquered the land, leaving the less evolved fishes behind. Dinosaurs are the presented as the grand losers of the evolutionary struggle, forever vanquished by superior mammals. Even so, fish represent half of all vertebrate species and dinosaurs survived for 130 million years, yet both still occupy the dispossessed underclass of natural history.

Welcome to the Church of Darwin. Follow our scriptures of logical meanderings and hidden sacred non-questions. Contemplate a universe that appears beautifully crafted, yet is evolved of random happenstance, astronomical coincidence and utilitarian necessity. Imagine that you inhabit a world devoid of imagination. Disregard your own mind and subvert your emotions, as our world has neither. Know that life is an endless battle of greedy machinery. Indulge in the pornographic idea of procreation as pure mechanism. Dissect the living until it's an empty corpse. Inter the soul in piles of data. Live in a city born of ideas and study nature born of randomly mutating genes. Take a leap of faith so great that you must deny the possibility of your own existence.

A long list of problems arises with the view that evolution occurs through random mutation and utilitarian necessity. The most obvious difficulty is that Darwinian evolution lacks an intelligent creative principle. It's an absurd notion that beauty, harmony and intelligence can be formed by blind mechanism. It's much simpler to believe that they are inherent aspects of evolution. Because science is based on a materialistic platform, it must deny or ignore the presence of mind in nature. Moreover mind cannot be measured or objectively proven to exist and therefore excluded from scientific theory.

Since Darwinian evolution focuses on the evolution of the individual it can't account for how things evolve as a whole. Neo-Darwinian theories overlook the connectedness of all life, portraying life as a constant battle to spread genes. In the popular book which is too irresistible not to mention, The Selfish Gene Richard Dawkins sees living beings as bags of chemicals blindly intent on replicating. He writes "remember we are picturing the animal as a robot survival machine with a pre-programmed computer controlling the muscles." This cold reductionist, ultra materialist view pervades much of science and serves to justify the pathological greed that infects our culture. It's techno-nihilist undertone is amoral and soul stealing, a world view of a cyborg in a post-apocoliptic wasteland. Incidentally, nowhere in his book does he explain how he, as a replicating bag of chemicals can think feel or write a book. The idea that evolution occurs through self- interest is not necessarily wrong, it is the idea of what the self is that is limiting. An expanded definition of the self would include the group, community and galaxy. All beings evolve simultaneously on every level. A biologist can view an oak tree as a selfish being evolved to crowd out other trees, suck up precious resources, and trying to take over the forest with it's millions of seeds. A more holistic viewer might see the oak as a life giver providing food, shade, housing, oxygen, cleansing and support. Besides it's utility to others the oak is a work of art, a wonder of creation flowering with ancient wisdom.

Any evolutionary theory that focuses solely on one level of reality would never be able to express the synchronous eruptions of the evolutionary process.

How can Darwinian Evolution explain the following?

* aging or death (not traits suitable for survival)
* Group selection (white flag-tail of deer, stripes on zebra)
* diversity instead of monopoly
* momentum toward healing and harmony instead of entropy
* sexuality instead of asexuality
* altruism and compassion
* Traits that evolve with foresight or goals (flight, oxygen sky)
* suicide (lacks survival value)
* homosexuality (not too likely to be passed on)
* rarity of rape among animals
* suicide
* birth control and abortion
* homeostasis of atmosphere and ocean

* rarity of poisonous creatures (shouldn't all beings be quilled and inedible?)
* conservatism of basic design (insects, tetrapods)
* rarity of tool use (obviously favorable for personal gain)
* restraint and ritualized aggression (bucks, scorpions)
* species that no longer physically evolve (horseshoe crab, rhinoceros)
* artistic expression (dance, music, beauty, drama)
* inefficiency of photosynthesis (reflection of green light)
* lack of transitional forms in the fossil record (Cambrian explosion)

Darwinian logic is a mind loop, a tautology that leads the thinker in ever an tightening snare. As the origin of species is based upon the survival of the fittest, we might then ask who are the fittest. The answer of course is those that survive. Using this skewed logic we are asked to look upon every attribute as a survival weapon, with the idea of utility being the only reason for being. The level of absurdity is raised as we consider the usefulness of a flamingos' pink or a scarabs' shine. Beauty and humor are outlawed extravagances of an unschooled sentimentality.

As Darwinian logic locks its loops on the social sciences, we are denied the aspects that complete our souls. As our motives are reduced to a gene spreading mechanism, we translate this prophecy into corporate efficiency, for all else all must be time-wasting frivolity. Freedom is deemed an illusion by a philosophical notion that seeks to erase the landscape of our imagination. The love that holds the world together is banished by a creation myth that envisions the world as a battle of zombie replicators.

Several other evolutionary theories have been proposed. V.C.Wynne-Edwards proposed a theory of group selection. Realizing that Darwin's' theory would create a runaway system that would destroy the food base, this theory says that evolution occurs at the level of the group as well as the individual. Mutationism is a theory proposing that a species originates due to a sudden jump in genetic change. Orthogenesis theorizes that beings evolve along limited pathways of development similar to those that occur from birth to old age. Lamarckism is based on the idea that traits acquired over the lifetime of the parents are inherited by the offspring. These theories are seen as competitors with Darwinian evolution, despite evidence supporting all of them. Each theory is a piece of the puzzle. These ideas also fit neatly with those of Rupert Sheldrake.

In A New Science of Life, Sheldrake theorized that fields exist in and around all of matter. He refers to them as morphic fields. Science has long accepted that fields exist. Gravitational and electromagnetic fields transcend the borders of objects and magically affect and order the surrounding matter. Iron filings dropped around a magnet create swirling lines of force as the hidden field is revealed. Fields serve as the underlying organizing system of nature. A morphic field is analogous to the spirit of a being. It contains a collective memory on which a being draws upon and contributes to. It is the evolving life-force that nurtures a plant from a seed and patterns an embryo into a baby. It is the attractor that sculpts matter into living forms. As a being develops it is attuned into the field of its species, drawing upon a pooled memory. Past forms influence present ones through morphic resonance- the influence of like upon like through space and time.

Sheldrake uses the example of crystals. When a new compound is created it takes a long time for it to crystallize, as no morphic field exists to pattern it. When it does finally take form a new species of crystal is created. Now the substance will easily crystallize toward the original type as a new field arises.

Organisms exist as a nested hierarchy of fields. Particles exist within atoms, within molecules, in cells, tissues, organs, organisms, communities, etc. Each level tunes in with it's own type and is influenced and regulated by other levels. A being develops by retracing the steps of its ancestors. Environmental, genetic and spiritual influences create new variations that then become part of the field of its species. These new varieties influence future generations, which develop along similar pathways and life histories. The more often a pathway is traced the more it is etched in time, and is more likely to be followed in the future. An organisms' individual memory is part of the memory of its species and universe. Something learned may become a habit and is passed down to offspring as intuition. Habits and ceremonies become preserved in the species as instinct, which help a creature deal with life circumstances that were experienced by its ancestors. Commonly traveled paths create a powerful gravitational force that limits free will. Creativity and play counter the rigidity of instinct and form new branches on old pathways. A destructive pattern may be prevented by breaking the cycle of rituals that initiate that particular pathway.

Morphic fields are not fixed, but change as a species evolves. The most common varieties exert the most pressure on future beings. The more alike two things are the more they resonate with each other. This is true of behavior as well as form. It is a tuning fork vibrating to a sound of the same wavelength, or a radio tuned to a particular frequency. Morphic resonance includes the dimension of time, as all things resonate with similar things that have existed in the past and possibly the future. Whenever we learn something we resonate with all of those beings that have traveled that path before. Learning to speak, drive, or program computers, becomes easy as we follow the path laid down by our ancestors. Developing into a human from a zygote or learning to play the piano occurs as part of the same process. Personal, cultural, biological and cosmological evolution are aspects of the grand unifying field of the universe.

Sheldrakes theory of morphic fields is limited by the materialistic paradigm that cripples and corrupts science. If matter was known as a subset of mind instead of its reverse, then everything falls into place. A morphic field is better known as an idea, its pathway is imprinted in Mind. We can intimately feel how evolution works, because it is the way of our thoughts. The creativity, memory, and organizing principles of our minds are an extension of those that have evolved our world. For people living in western technological cultures this may be hard to see. We often emphasize conscious control and engineering over organic freestyle creativity. It is the difference between a city and a forest, rationality and art, work and play, left brain and right. Both are necessary aspects of thought and evolution. Whether it's a city or a forest, they both represent the materialization of the mental processes that dreamed them into existence and are forever intimately connected. Many artifacts of civilization appear to us as essentially different then those formed of wild nature. Civilized objects are usually an assortment of parts combined together. Wild objects are often grown from the inside out. Each one represents the applied philosophy of the mind behind them. We can accept the idea that an intelligent being can create an object outside of itself. We believe that human beings can engineer a car or even that God can create a world that he stands outside of, but we can't easily comprehend an intelligence that creates itself from the inside out. We therefore don't understand the workings of our own bodies and minds.

"How do the weak gluons cause particles like hadrons to disintegrate. The other gluons, the photon and the graviton we previously described do not do that. Why are weak gluons so special? To understand how a hadron can disintegrate we have to remember what a hadron is made of –the different flavor quarks up, down, strange, charmed, bottom and top. What the weak gluons do is change the flavor of the quarks. For example a strange quark by interacting with a weak gluon can be converted into an up or down quark. This means hadrons possessing a strange quark can change into hadrons with only up or down quarks."

From The Cosmic Code
By Heinz R. Pagels

Quantum physics is the science of transcending science. At its heart it studies the sublimation of Matter to Mind and its inverse. Early experiments showed that things are altered in the process of being observed. This destroyed the pretense of scientific objectivity and has ushered in a process that has yet to be fully realized, the overthrow of materialism. Humanity has been shielded from the implications of solid objects reduced to wave probabilities. Where were the headlines that proclaimed "Science Proves Magic is Real" following discoveries of particle entanglement and teleportation? Few sane people outside if physics acknowledge that objects can suddenly appear into empty space only to disappear an instant later. Even fewer discuss such irrationalities as gluons, neutrinos and antimatter.

Even as thousands of new technologies are unleashed on us hardly anyone investigates the mysterious influences behind their functioning. These forces defy reason, thou they are synchronous with obscure mathematics. An assumption is made the science behind them is understood, but there is no requirement for unreasonable forces to be knowable for them to be utilized, just repeatable. A scientist might not have a clear idea what causes gravity, but this doesn't prevent him from floating away. Even simple technologies are often based on very strange events. Electricity can pass between two touching wires due to what is called quantum tunneling. When this occurs a certain amount of electrons in one wire are able to dematerialize and appear in the other to complete a circuit.

Science is placed in a great quandary as physics has become heretical to its positivist doctrines. An immense pyramid of data stands on a rapidly dissolving foundation. A virtual quarantine has been placed around quantum mechanics to prevent the contagion of mysticism from spreading. Perhaps like the most ancient heresies this system of knowledge is not taught to the masses or even fellow scientists in order to preserve the hierarchy of control.

Within the revelations of the new alchemy is the Inverse Quantum Zeno Effect. It is defined as a series of dense measurements of a quantum system along a particular path that will force the dynamics of the system to evolve along that path. In other language, matter which is solidified by being observed can be channeled towards a specific direction by being obsessively measured over time. What is revealed is a secret means by which science engineers the world. Long before this effect was discovered its technique was already standard practice. Legions of observers descended onto the earth as if from another planet, separative viewers trained to objectively study every phenomenon. Missionaries of practiced detachment sat amidst the earths' tribes, and reduced them to doctoral dissertations of ethnology. Tenured grave robbers looted ancient crypts to make sure the dead will not rise. Armies of voyeurs with notepads peaked into bear dens and birds nests in order to systemize their behaviors. From hydrothermal vents to Antarctic icepacks the planet was reduced to a frontier observation post. Eventually satellites, microphones, scanners and digicams will stand watch over every last body, landscape and culture and saturate the solar system with an invasive duality.

Honestly believing that we were seeing things for how they were, us measurers failed to see how our presence radically changed all that was placed before us. Trained by the academies of the enlightenment to stand outside of what we were observing, we cracked the world in two. As all that is persistently watched becomes a frozen shell of what it once was, we measurers formed the idea of a physical reality, a graveyard scraped out of the enchanted realm of possibilities. We discovered the laws of nature then we set out to enforce them to prevent our waking life from reverting to dream. In the neutral gaze of our instruments the ghosts of our primitive pasts failed to register their presence and were thankfully declared a delusion. We believed in the noble deed we were accomplishing. We shed light on parts of existence that have evaded our consciousness. In so doing we believed that the broken world could be sewn back together through our interwoven thoughts. Amputees lost in the shockwaves of the Fall could return and bask in the light of our awareness.

It has been said that it is better to light a candle then to curse the darkness. Driven by this tricky logic we set about lighting up the world. A war against the darkness was instigated and streetlights lit up the rapidly retreating night. Just as a child is given a nightlight to prevent the monsters from morphing out of the black, our habitats were illuminated to keep out the haunting presence of the unknown. As it is with any war, the object of our persecution was multiplied through our attacks on it. A million candles sat burning through the night and a million shadows were cast where the demons waited and plotted their revenge.

Maybe what we need is a science of science. Clearly it would be steeped in irony as the researchers of this new field would be in a position to study themselves studying. A question emerges.... Would an infinite progression of self-conscious scientist reveal the face of God.

To obsessively stand outside oneself and feign separation is an act of derangement, it is living as if to engineer against integrity. It is to deny that in the end the only object that we can know to exist is the Mirror. The underlying pure state of being can't be reached through separation and therefore a deep knowledge of existence can only be attained through communion not observation. Reality when spied upon becomes self conscious to the point of becoming a simulation of itself. Detached from the eternal life-force, creative evolution is halted and beings become mechanized caricatures of what they truly are. The massive surveillance system that has been erected on this planet serves its purpose to alienate life from itself by suppressing coherence.

The membrane between mind and matter is the quantum realm. It is composed of quarks riding probability waves, flashing in and out of existence. It is a permeable layer of thought quanta. The intricately infinite patterns formed by the waves, paint all the beliefs and perceptions of the universal dream. Here inhabits the possible. Every thought and vision pulse in the ocean of our deeper selves. It is the dreamtime flowing from the realm of the timeless, slowly morphing into gasses, then liquid and finally congealing into our solid world. What determines its materialization is the strength of the field behind it. Fields are energized by our conscious observations and subliminal beliefs. All things and beings are alive, their fields being their airy spirits. Matter is the iron filings attracted to the energy emanating from the magnet of our soul. It is the condensation of our thoughts.

Within the field is the memory of all that we once were. Our genes and brains are not the source of memory. Instead they are the portals through which our ancestral pathways enters. Brains and DNA are narrowing devices as they reduce the infinite into the particular. Through them we are shaped and through them we shape our world into the image of our wishes and fears.

What is considered to be the conscious mind is the least conscious form of thought. It is the illumination of a particular object and the subsequent neglect of the context in which the object inhabits. It is the mind of separation, one ego communicating with another. What is unfortunately called the unconscious is the higher state of mind. A better term might be meta-conscious, as it is Mind working on all levels at once. It seems unconscious to modern humans because of the barriers we have constructed against our true selves. We have been taught to believe that communication is the greatest sign of intelligence, neglecting communion, which is the truer and deeper state.

There were once within me scientists and mathematicians put to work in giving structure to all that was evading the drowning grasp of my mind. They were entrusted as gatekeepers of solid reality. Yet I was told that an atom fired at two cracks in a wall dissolves into numbers, enters both holes at once and reconstructs itself on the other side. The high priests impotence was revealed in confused doctrines of uncertainty and undecidability. Isn't it true that both light and logic equally dissolve into the singularity of a black hole?

Stripped of the pretense of knowledge I reached the heavily guarded boundaries of organized thought and consensual reality. A conditioned fear of madness became the final obstacle in transcending illusion. They once told me that only the insane can inhabit a place without boundaries for entropy is inevitable. Their despair at coming to that conclusion told me more then their insistence that psychosis is anarchy of the mind. Breaking the restraints was all that was left to do before entering utter awareness.

Light and mind have an intertwined history. The changing view of light corresponds to major changes in what is believed to be real. Even hardened rationalists must admit the existence of an inner light within mind. This inner brightness illuminates the outer reality as well as our innermost thoughts. If you close your eyes and think of a firefly—suddenly a flash of light. Many ancients and tribal peoples believed that a ray of light emanates from the eye. The mathematician Euclid believed that if you drop a needle on the ground a ray beams forth from the eye, scans the ground, and only when the ray strikes the needle can we actually see it. As fantastic as it sounds, this view of light was widely believed up until the last few hundred years. Plato writes "the fire of the eye issues a gentle light and coalesces with the daylight". This marriage of inner and outer light is key to Goethe's view of vision: "the eye is formed by the light, for the light, so that the inner light may meet the outer." When this theory is considered possible, it becomes a marvelous way of knowing what consciousness is. A directing of light onto concepts, ideas, forms, memories, dreams as well as into the inter-subjective material world, through the portal in space-time we refer to as the eye.

We're taught to disregard the most vital elements of existence. Encouraged to see things head-on, we ignore the periphery. Our focused glare freezes subtle vibrations into the matrix of hardened matter. Meanwhile the real action occurs at the corners of our eyes and the limits of our vision. We insist on our photographs to be clear and sharp edged, cleaning up the phantoms through computer enhancement. We may acknowledge a photos' affect on our emotions, but we rarely ask how it can be that ones inner life can be reflected on paper. The light reflected by an object is only half the story. What of the light of consciousness beamed from the mind and eye of the photographer. On film, the combined imagination of seer and seen is integrated and distilled yet we often fail to wonder.

Soul is lost in our search for clarity. Records evolve into cassettes soon to be Dolby enhanced, finally their fuzzy analog complexity is lasered into binary slices of prismatic plastic. The infinite resolution of film is reduced to magnetized metal, digitalized on electro-plated tape and further clarified into DVD. We fail to ask what is lost when our snowy screens give way to plasma sharpness.

The inner Scientist demands the repeatable at the expense of the anecdotal, it insists on the measurable thereby obscuring the eternal. Meanwhile, life occurs at the bleary edges and recoils at the dissecting harshness of our glare.

The world was but a stage performance. The attentions of science was not in the dramatic actions no matter how lavishly set or passionately acted, but rather with the calculating view of the stage engineer.

Bernard de Fontenelle, 1686

A biochemical view of mind has come to rule psychology. Most mental problems have been designated as chemical imbalances. This belief arises from theories that claim that mind is created by matter. This belief is widespread even though there is much evidence that our brain chemistry and neuronal connections evolve according to the pattern of our thoughts and emotions. Personality and chemistry are a continuation of the same process, each playing off the other. Only in death when the field no longer unifies the body, are they more easily separable.

Reducing mind to brain chemistry is a simplistic remnant of Newtonian thought. Quantum models come much closer to figuring out the brain. Matter is seen as vibrations patterned by energy fields. Vibrating atoms emit photons, like little transmitters. As more energy is pumped in the atoms vibrates in unison, creating the more ordered forms of matter. The formation of this condensate makes visible the underlying thought patterns. This coherent state shapes the matter into the likeness of its thoughts. Just as when patterns within wind are revealed in blowing leaves, thought becomes clothed and visible by matter. These coherent states of matter have been found in the brain. Electrical stimulation of neurons serves to synchronize nerve membranes into condensed phases of consciousness. Here the spirit interfaces with the brain, wiring it to a quantum computer that matches the rhythms and thoughts of its user. Like nerve membranes, DNA can also vibrate in unison. Here the vital field can play the rungs of DNA like keys on a piano, proteins are the music that flows out of them. All the DNA of the body becomes united, building and healing tissues that evolve towards the ever-changing form of the vital field. The ability of matter to align itself as one is known as the pure state, here solid objects dematerialize into a superposition of all possible states. It is the science behind superconductors, lasers and condensates. In the ultimate sense the Pure State is called the Singularity and to experience it is to reach ecstasy. The universe can be acknowledged as an immense organism, breathing in and out, towards and away from its source, with different aspects ever vibrating with coherence and consciousness. Scientists searching for the unifying theory of all forces have the answer in Mind.

Buried within Darwin's theory of sexual selection, are the seeds for a much larger, and inclusive evolutionary theory. Watching birds, mammals and even flowering plants, it becomes clear that within the natural world there's a great striving towards artistic beauty. Flowers attract insects as well as humans by their fantastic forms, fragrances and colors. Bower birds collect colorful shells, flowers and fruits, building spectacular displays to attract mates. Springtime becomes a symphony of sound, beings as diverse as birds and frogs fill the air with their song. Courting cranes dance with the grace that ballet dancers can only approximate. Every form of artistic creation, in an infinite variety of mediums is practiced. But why? These performances run counter to the theory of natural selection, which would quickly eliminate traits not designed for utility and survival. Why sing or dance when it expends such energy? Why grow beautiful feathers or iridescent scales when it would make you so vulnerable to predators. Darwin answered these questions with his theory of sexual selection. Basically saying that these beings express their beauty in order to attract mates. These traits are then more likely to be passed on to offspring. But in order for this hypothesis to be true one must admit that animals have an aesthetic sense, something most scientists are unwilling to admit. For to admit it would be to say that nature is an artist, as well as a technician. It would have to concede that evolution moves not out of necessity, but from an inner urge towards the irrational impetus of grace, an indefinable quality. What sexual selection recognizes is the close relationship between creation and procreation. Sex isn't as much a logical endeavor, but the ultimate expression of universal creativity.

The idea of Eros has been reduced to sexual desire. In it's greater sense it is universal desire. Sex is the intensification of the evolutionary process, the literal flowering of existence. The idea that a being would choose its mate for artistic reasons flies in the face of those that believe that the "fittest" are necessarily the most aggressive or greedy. What this implies in our war torn and overexploited world is self-evident.

Evolutionary theory is the creation mythology of our age. It is a paradigm that has barely entered the consciousness of humanity, except in the crudest form. What we believe about nature's evolution we apply to our cultural evolution. To neglect the artistic and spiritual aspects of nature we atrophy those qualities in ourselves, for how can they evolve in a vacuum? Denying the Artist, soon becomes imprisoning the conscience and finally leading to moral collapse and dissociation. The evolutionary trend towards greater beauty is an outgrowth of nature's highest moral lesson, that of creating harmony.

Eros is the creative impulse. It is the selector enacting the divine will and the universal motivation of evolution. Eros is also described as a yearning, a desire for what's missing or as Paul Tillich describes it, the "unreflected striving for what is noble." Eros is not just the act of making a choice, but choice based on love. Love can be seen as the ideal evolutionary pathway in which Gaia oscillates. This pathway is analogous to the Tao, the way, the good, grace or any number of ineffable concepts that represent this ideal. The goal of many eastern religions and mystics is to align individual will with the Divine and Universal; to become one with Nature. Choice that is made counter to this principle produces imbalance, incoherence, evil, uniformity, linearity and the fall of man.

In a world Made dependant on biotechnology, gene splicing and genetic testing many assume that genetics is a subject well known to science. A popular misconception is that everything exists in the genes. The gene has become an icon. One hears talks of genes for alcoholism, obesity, height, aging, intelligence, breast cancer, insanity, violence and any other possible traits. The public has accepted this image as truth. But as far as science knows the genes only encode information to build amino acids to create proteins. No manic depression genes, no hyper sexuality or gambling genes, just proteins. There is a major assumption that all inherited characteristics are stored in genes.

Even the science of biological inheritance is called genetics. The DNA is often referred to as the blueprint of the body. Where is the evidence? Nowhere in the genome can you find a blueprint for the shape of the hand, the formation of kidneys, or the song of a sparrow. The genes aren't even a blueprint for creating a protein, instead they create a template for the RNA molecule to assemble proteins in a ribosome.

Some traits are directly caused by the existence of a gene. The color of a flower might be created by a specific protein and therefore dependent on a specific gene. Many diseases are caused by a mutation in the genome, creating a faulty protein. All traits to some degree are affected by the body's genetic makeup, as proteins are intricately linked with every aspect of physiology and behavior. Just as Darwin's idea of natural selection is a part of evolution, genetics is an aspect of inheritance. They are both based on models of nature that are mechanical and utilitarian. The result of seeing the body as a mindless automaton is to select processes that fit into that particular paradigm and to disregard everything else. Genetics is based on the same philosophic framework that supports Newtonian physics. This framework helps describe a genuine pattern that structures matter, but it is a limited truth posing as an absolute.

The idea that DNA is the control center for the body is a rationalization for centralized power structures in our culture. Genes are then seen as the blind dictators of determinism, manipulating the body and mind towards maximum metastasy.

The mechanistic framework of biology struggles to comprehend how the staggeringly complex universe of our bodies can be encoded in 30,000 genes. Chimpanzees share over 99% of our genes. Could a few hundred genes possibly explain the difference between chimps and humans? Every bone, nerve and muscle has been reworked including massive changes to the pelvis, cranium and vocal chords. Furthermore, if the genes were the body's' blueprint it would make sense that simpler beings would have much fewer genes. This is not the case. A mustard plant only has only 8000 fewer genes then we do. The nematode worm has 20,000 genes in a body containing just 959 cells. The human body in comparison has over 100,000,000,000,000 cells differentiated into hundreds of types growing in massively complex tissues. The amount of DNA also seems unrelated to a creatures' complexity. A humble amoeba has a genome 320 times larger then you do. Some plants have thousands of chromosomes. Something doesn't add up. Insisting that memory is stored digitally in DNA has been a stumbling block for science. One possible explanation is that the genes act as initiators of pathways, rather then carriers of information. The DNA would act like a radio receiver, translating data infused in the morphic field. Genes are switched on and off by the silent intelligence that underlies the body and all of matter. The timing of genetic transcription involves the interplay of all of our evolutionary history with the needs of the present moment. The body needs to evolve at all levels at once. Imagine the complexity of healing a broken leg. Lymph, blood, nerve and bone tissue needs to be activated in precise and integrative ways. All of the body's organs must collaborate and act in unison. Billions of cells produce the necessary proteins to realign the body towards its idealized healthy image contained in the field. Healing a broken bone is easy compared to creating an entire body from a single cell, yet this miracle is a common occurrence. No amount of stored data could hold enough information to monitor and maneuver every molecule of the body to deal with chaotically changing conditions.

Many genetic studies have been performed on identical twins, as their genes are nearly the same. Twins, besides looking the same often have similarities in almost every aspect of their lives. Even twins raised apart in radically different house-holds have an uncanny correlation between events in their lives. Sometimes marrying spouses with the same name or giving birth on the same day. This seems to point to a strict genetic determinism that guides peoples' lives, but there is another way to look at it. Because of their shared DNA, the lives of twins highly resonate with each other. They have a shared spirit that weaves their lives together even more so then with their parents or lovers. Twins evolve in similar ways not because they have the same programming but because their development revolves around a shared soul.

In dogs the slightest change in DNA means the difference between Sharpeis and shih-tzus, huskies and Pomeranians. The entire physiology of a breed might be drastically different from another, yet their genes may be 99% identical. This gives us great clues to the higher functions of the genetic code. The DNA seems to act as an initiator sequence. Minor changes in the order of base pairs can lead to major somatic change because it would direct the vital field towards a different morphological pathway. One change in one base pair of the DNA ladder creates a new amino acid. The field responds by shifting its growth patterns to incorporate this altered gene. Ingrained in the field are all of the ancestral pathways. These exist as embryological remnants in the personal evolution of each dog. The altered gene sequence changes the vibration of the DNA. This vibration then resonates with that of an ancestral being that had a similar genetic sequence. The wisdom inherent in the vital field initiates the dog towards a viable growth pattern. The pattern of an ancestor that was able to survive with a similar gene sequence. If no viable ancestral pathway resonates with the new sequence the dog is likely to die off in its early stages of growth. It may also create a new pathway, forming new traits, or even a new species.

Breeders select and refine traits through their selection. Each breed of dog a continuation of an ancestral pathway. The humans act as transcendent gods guiding creation toward specific directions. Natural evolution is a similar process, except what guides the breeding process is more mysterious and less controlled. Each being has a choice as to which mate is selected. Other factors may include chance, love, fate and hidden intelligences that pattern reality. The causes of mutations in DNA are equally mysterious. The forces that guide evolution are as unknowable as the Great Mystery and as near to us as our own minds.

Evolution can occur at a fast pace because it can utilize any idea or design that has already been made. Art, technology and domestication are an extension of this process as they use old ideas as a framework for creating new forms. The birth of totally new ideas is much more mystical, as they seem to erupt into consciousness from out of the void. It's relatively easy to engineer new designs using a known mathematical structure, but imagine trying to invent the actual mathematics. The concept fringes on a great puzzle-How can something be created from nothing? Most of the greatest human mathematicians, explorers and scientists weren't as much creators as they were rediscoverers. The concept, continents and paradigms that they found already existed prior to their human discovery. What is mind- boggingly bizarre is that if such initial creations such as the speed of light were slightly different, it's questionable whether anything at all could exist. It seems that all original creations and archetypes must exist outside of time. In this way they seem to pre-suppose the universe that is based on them. Any creator must travel deep into the eternal and experience the grand unity, then return back into the realm of time and infuse existence with the divine presence. The solution of any problem and the answer to any question is found at the center of being, and defies concepts such as cause and effect. Plato considered this realm as the perfect, ideal, and eternal, but its reality seems meaningless without the existence of the fractured, fleeting and deficient. The broken symmetry of our world is what makes it so fascinating. The perfect can only evolve when disguised as the flawed.

The evolution of bacteria can be directly studied in a petri dish. The use of antibiotics creates new strains resistant to these chemicals. Many of the new antibiotics are synthetic and haven't previously existed. The question is how could the bacteria evolve resistance to a chemical it has never experienced? Most biologists would say that a chance mutation in the bacteria's DNA selected a single germ for survival, which then formed a resistant strain. This may be the case except that resistance seems to occur for every new antibiotic created. Certain strains of TB have become immune to seven different types of antibiotic. Defying current models of natural selection highly resistant strains arise during a required multidrug therapy. A single mutation in this case lacks a selective advantage. How could it be that by pure chance there is a bacteria already evolved to survive any possible poisons in various combinations? A simpler solution would be to theorize that bacteria have an intelligent creative mind, and when confronted with a problem, they solve it. The key to problem solving is imagination. Virtual states of being are endlessly created as the particle reveals itself as a wave of probabilities. Imagine a bacterial colony living in a drop of milk. As energy sources are exhausted the colony senses the lactose surrounding it, but is unable to utilize it as it lacks the enzyme to digest it. Confronted with this dilemma the colonies DNA searches possible futures as it dissolves into the unified field. Using billions of years of ingrained wisdom and virtual trial and error a solution is reached as the germs reach a coherent state. The problem is an easy one, all that is needed is a single point mutation in its DNA. Given enough energy the mutation crystallizes out of the virtual realm and reconfigures the genes in the colonies offspring. In hundreds of experiments performed since the controversial Cairns paper in 1988, bacterial, yeast and animal cells have shown much higher rates of adaptive mutations then chance could ever allow.

Bacteria are the ultimate biochemists, able to create millions of different complex molecules within their own bodies and to trade information amongst each other. Science uses two different models to describe how a chemical evolves, one for the chemist applying his intent to come up with solutions, the other for the bacteria, using pure chance. It strains the mind to think that man uses intelligence to make decisions and the rest of the universe uses a blind "natural selection." It's also a violation of the principle of Occams Razor, which always seeks the simplest solution. A world built of the random accumulation of disparate pieces is an idea based in superstition. The astronomical coincidences involved require faith in the most improbable of scenarios. It also provokes the notion of a supernatural creator rather then mind immanent in nature that we know must exist if we are to acknowledge our own awareness.

Beings evolving under the same evolutionary pressures, flow towards a common attractor, and are possessed by a unifying spirit. This accounts for the ability of many animals to mimic others. Many stick insects and katydids have evolved such perfect camouflage that they duplicate twigs and leaves right down to the veins, tears and fungal growths. Creatures that seem to have evolved independent of each other have produced remarkably similar body types, coloration, and behaviors. Called convergent evolution, this process allows accumulated wisdom and memory to solve similar evolutionary problems. With the evolution of the fish, a body type was formed out of an infinite array of swimming creatures that would later give rise to amphibians, birds and mammals. The fish form was also faithfully reproduced in reptiles such as ichthyosaurs and later in mammals such as dolphins. Once this form was created it set a precedent that once learned, was later reapplied by nature as needed to fill niches.

Just as the fish design is applied to other swimming forms, flying (which is swimming through air) shows a similar phenomenon. Probably originating in insects, flying arose separately in fish, reptiles, birds and mammals. All show similarity of design, with a "t" shape, similar balance and attachment points of wings. The evolution of flight is a stumbling block for current evolutionary theory, especially because it has evolved so many times. If evolution occurs without mind and foresight, and in small stages, how could something as complex as a wing evolve, as flight is an all or nothing venture. A partially formed wing would get you nowhere. To evolve flight, the entire body of the creature needs to be reworked. Birds can take off because of their aero- dynamic shape and super light- weight resulting from hollow feathers and bones. The immense complexity in making a ground animal fly suggests foresight in design. This is not to say that evolution always tends towards known goals, but that it also relies on the serendipity of chance and chaos. Penguins, which have evolved to swim, had ancestors that had evolved to fly, that had ancestors that evolved to crawl, who had ancestors that evolved to swim.

The convergent evolution of flying animals converges further based on the niches they need to fill. The sphinx moth and the humming bird both drink nectar and look and act identical in flight. Bats and cave swifts both come out of crevices and caves at dusk to eat insects. Their look is similar, and both use sound as a second sight.

Separate continents have formed very similar types of plants and animals, despite distances of space and time. Each continent is a world in itself, yet familiar forms independently emerge displaying the unity of mind on earth. An anteater in South America shares its spirit with the aardvark in Africa, the pangolin in Asia and the echidna in Australia despite different ancestry. Taxonomic categories, which are based on lines of descent, are only one way to group beings together. Another way would be based on homology and similarity of spirit, regardless of physical ancestry. Many native peoples regard themselves as descendants of various plants and animals. They recognize their spiritual ancestry through identifying with their totem creatures.

Islands that have no mammals such as New Zealand, often have birds filling mammal niches, such as the flightless and hairy kiwi. Almost any known creature has had an analog that has lived a similar lifestyle. Even oddities, such as platypus fill a niche somewhere between otters and diving ducks and have the body parts to prove it. Many biologists try to explain away convergent evolution as the logical outcome of animals evolving under similar conditions. This overlooks the fact that infinite pathways and forms could exist. If evolution were a blind process, diverging sequences should produce infinitely different creatures. Why should a triceratops be like a rhinoceros, if given an infinity of forms?

The conservancy and repetition of form in nature, give clues to the habit and character of our universe. The spiral, the face, the eye and the tree are signature forms and symbols, revealing bits and pieces of the great mystery that imagined them into existence.

There is a remarkable correlation of many corals and fungi. The fanciful shapes and colors of reef beings are echoed in coral mushrooms and slime molds. Mushrooms also grow into highly phallic forms, revealing aspects of their underlying spirit. Flowers also display a very recognizable correlation with mammalian sexual parts, as they too are reproductive parts. The convoluted "brain" form grows in reefs, forests, vegetables and within our own heads, any place that surface area is of high importance. The fractal complexity of a seahorse's tail resonates with the grace of an unfurling fern and a crystal laden snowflake. These forms are produced with a seemingly simple fractal geometry; others are so complicated that a repetition by chance would be unlikely. The form of the face has appeared in nature in many diverse forms for hundreds of millions of years. This preference for a head, two eyes, snout and mouth is evident in insects, spiders, centipedes, fish, amphibians, mammals and birds. It is also revealed on butterfly wings, crab stomachs, aspen sprouts and birch trunks.

Insects having, evolved four hundred millions years ago still grow in three segments, have six legs, wings and antennae. This conservation of form remains despite their ability to evolve rapidly. Sometime before the invention of the insect the earth reveals in its fossils a creative outburst of body types in what is known as the Cambrian explosion. Defying Darwinian gradualism, all twenty five to thirty phyla of animals that exist today evolved around 560 million years ago. Many other unique and wild looking animal fossils found in the Cambrian rocks can't even be assigned to any living group. The Cambrian explosion is the meeting point between Darwinism and Creationism, a sudden genesis of animal life, yet having the continuity and bloodline of all of the earth's beings. A similar burst of diversity bloomed four billion years ago as bacteria evolved a huge selection of lifestyles and forms of metabolism and reproduction. Two billion years later with the creation of sex evolution, in a stroke of genius created the pathways to a multi cellular being with a scale of existence and complexity familiar to the naked eye. These biological revolutions and artistic movements didn't destroy the old ways, but used them as foundations to create even more diversity and artistry. Each step, from quarks, to carbon to aardvarks to galactic clusters evolves as a tightly interwoven collage of divine fabric with fibers stitched through inconceivable dimensionality.

The past never passes, but instead is ingrained into the essence of matter as memory. Time is made of the substance of memory, funneled from the infinite past and future into the matrix of matter through a cyclonic vortex we call the present. The future is that which is out of reach of the senses, the past exists within things. The young sapling hides buried in the concentric concretions of time within the ancient redwood. Consciousness is the rapidly dividing tissue of the present, points of light scanning the horizon of possibilities towards new growth. What exists today is the condensation and crystallization of past thoughts and actions. A nutritious upwelling of wisdom springs from the buried, but ever living past. Every conception flashes the sequence of thought that brought it to that point. Our own personal evolution of being, traces life from the beginning of sex through cell division, spherical forms, worm-like embryos, fish complete with gills, monstrous amphibians, furry mouse-type creatures to human babies. After that, the horizon points the way and the past makes suggestions, but we begin shaping ourselves. This is accomplished through that which we take in. Food and thought are parallel concepts, as they are the substances that compose our being. It is humbling to think that we were once a slug like sac of undifferentiated cells, budding arms like a hydra and sprouting colorful organs beneath a translucent skin. Our entire past tantalizingly reveals itself as the sum of our travels. Evolutionary time warps may birth human babies with tails, fur, scales, snouts and gills. Sometime there emerges from our shadowy past, giants and dwarfs. Oddities such as horns, hooves and claws may result from converging archetypes of totemic spirit. They are often the result of genetic abnormalities redirecting growth pathways towards distant types. Downs Syndrome often referred to as a genetic defect creates a novel form of humanity. Far from being a defect, these types are often a gift to those that get to know them. Our paths are many times deflected by gifts and traumas that channel evolution towards archetypes. On occasion, we are thrown so far from the route that we've been following, we have no choice but to wander into the unknown depths. Once the path is laid down others may drift or direct themselves towards it.

We are all that we once were. The pizza delivery man we were at 22, the shy Child of seven and the gibbon 10 million years ago are held within, waiting to be expressed. Those incarnations nearest in time carry the most momentum into the future. Moments of emotional impact have a strong gravitational pull on our psyches. Powerful emotions strike deep and echo shockwaves from the past into the future and back again, an earthquake pulsing tsunamis across dimensions. Lives ride each wave towards the epicenter, and follow them out after the impact. Each crest a synchronous echo of the big event, until the sea finally smoothes over and only a ripple remains. With each shakeup evolution staggers, sometimes paths are diverted, sometimes new ones are created. Eventually all things move towards healing as paths are retraced towards their source. Scorched fields will one day bloom as fresh earth is revealed to the sun from the darkest depths.

Love and ecstasy flow through the mind and body, mending our fractured souls, diseased bodies and battered planet. Glimpses of unity and eternity align our being with the sacred flow. No reasoning or reading can bring this to us. It can't be learned or explained. How can one speak of a state of Mind, where all knowledge is known and all division an illusion. Priests of science and religion are placed as dictators, instilling partial revelations to spite the whole and criminalizing actions that would cure our blindness. We must become natives on a vision quest letting the wilds reveal our paths. The Absolute can be known only by becoming it. Portals of mind open before us, dams are breeched and dissolved, atrophied animals escape from their darkened industrial cages and into the sun. The alienation and loneliness of our ways are erased as the leaf acknowledges that it is also a tree. Hatred of others no longer is justified when the other is within. Tattered fragments and disenfranchised stragglers are welcomed back into the fold.

The seemingly sleeping ocean gleams with consciousness at a trillion points at any given instant. Flashes of illuminated phosphorescence infuse the depths with thought and action. The tree of life digs its roots into the water table, levitating sweet sap to invigorate matter with imagination. The ocean is the ultimate symbol of unity, a place where all rivers meet and mix. The birthplace of rain which drenches dry continents with lush and beautiful growth. To float peacefully at sea is to transcend the cage of the ego, as the soul returns to its origin of saline weightlessness and womb bound serenity. The sea both frightens us and draws us near, ready to redeem or dissolve us. All beings form membranes to prevent desiccation or dissolution, yet remain open to the oceans' currents and nutrients. We're taught to build levees and seawalls to insulate ourselves from the unknowable sea, till we lose touch with our source and our homes become lonesome islands. Once when we were free, we were encouraged to embark on voyages, seeking existential union as a source of spiritual sustenance. Now fearing outbreaks of freedom our wardens legislate against enlightenment, creating a climate where even a child day-dreaming is a subversive activity requiring medication to control.

Membranes of varying permeability insulate all living patterns, from belief systems to paramecium. They serve to protect a beings inner integrity from outside forces. Health and balance are a result of this inner coherence when it is in tune with the growth of the whole. Within this membrane exists the personal unconscious, the portion of the collective that is absorbed through heredity or experience. We are the genius loci of this personal ecosystem, the emergent mind of our inner habitat. Within it are forests, cities, rivers, animals and people, interacting with each other and striving towards coherence as they create ourselves. Our inner world is a real and imagined place, the habitat of our dreams. At it's depth it is rooted in the collective Mind, which helps shape it. All that we perceive moves into our internal landscape. Some things fit right in others cause conflict until a new balance is achieved. Just as in all natural systems, stability is dependant on diversity and integration of species. The key to healing body, mind, community and planet is the same. Psychology and ecology are equivalent concepts differing mainly in scale.

The boundaries between an organism and a colony forever remain unclear. The debate is dissolved in the mystifying waters of paradox.

A cnidarian is an organism that can exist both freely and as a colony. They have alternating generations that include an immobile polyp stage. The polyp buds miniature clones that are then set free into the ocean to live as nomadic jellyfish. These jellyfish mate, giving rise to the next generation of polyps. The polyps live and breed as one, building their homes on the skeleton stones of their ancestors.

The Portuguese man of war is a cnidarian that appears to the untrained eye as a jellyfish. On closer examination one finds that it is a vast colony of beings. Its float is composed of jellyfish and its gangly tentacles are a mass of polyps that are highly specialized. Some polyps capture food, others digest it, and still others live to reproduce. Somehow these colonies move and function as one. This type of alchemy is what holds bodies and worlds together.

Certain free-swimming amoebas can mass together into a slug type creature. The newly formed slime mold can then creep around and eventually sprout fruiting bodies to reproduce. Ant colonies split by a steel plate have been shown to match each side perfectly. A collective mind forms and organizes beings into a greater body. The larger is the deeper mind of the smaller. The created mind field doesn't exist compartmentalized but is spread throughout and is felt from within.

The ability to form a group mind isn't a trait confined within the bounds of a species, but occurs wherever a crowd becomes interconnected. As the roots of a forest reach out and intertwine, the trees begin to breathe as one.

As rain drips down into the tight rosette of a jungle bromeliad, a world in miniature is born. Leaves and bark drop in and decay. Birds and mammals stop by to drink, leaving behind their nitrogen rich excretions. The water soon fills with life, as protozoan, mosquitoes, and dragonfly larva emerge. Brilliantly colored frogs spawn in this new born world. Crabs, salamanders, slugs, worms, beetles and snakes visit its shores. The water magically stabilizes itself as bacteria work to bring it to homeostasis. The bromeliad thrives, nourished by the life it nurtures. It is the Earth and its water the Sea.

Balance isn't something that needs to be achieved, it occurs on it's own when all controls are lifted. The more we attempt to control anything or anyone the more fragments break away. Imagine trying to balance the body's physiology through conscious control and engineering. A network of supercomputers would be needed to monitor the functions of a single cell. To integrate the physiology of all of the tissues and organs would be an impossibility, coherence would quickly break down resulting in sickness and death. Modern medicine and psychology too often try to control the body and mind through surgery and medication, rather then to acknowledge that healing is a miraculous universal force that can be assisted but never engineered. The same problem arises with those that believe that the earth must be managed. This outlook is rampant among dam builders, forestry experts, factory farmers as well as with backyard landscapers and exterminators. A thriving chemical-pharmaceutical industry depends on this faulty belief system, peddling encapsulated imbalance in the form of pesticides, antibiotics, and anti-depressants. They further profit by selling treatments for all of the poisons and diseases they've unleashed. Schools have become institutions of thought control, a form of game management for the inner world of children.

Elimination of diversity and mind fragmentation are their consequences. Biotech companies seem intent on re-engineering and patenting all of the life forms on this planet. The underlying idea feeding these repressive practices, is that our bodies and earth are machines and our minds are computers. Components must be altered and tinkered with until everything runs like clockwork. Science and medicine stubbornly refuse to believe that a living creative process exists, it is a refusal to believe in our own nature. Indoctrination of these beliefs in the universities ensures that those in power embrace the ideals of control. These values structure the inner landscape with walls and fences, chained and tortured animals, wiring, gears and microchips, Frankenstien, and aliens, atom bombs and apocalypse. An inner police force and a hierarchy of judges enforce life-choking laws and soul robbing self-control, in a grotesque reflection of outer oppression. Outer conflicts are mirrored by inner battles. Our war on nature is a rape of the soul. The inner machine that is installed within us clear cuts our sacred places and pollutes our waters. It diverts our energy in order to mass-produce modules of dissonance, disease and dissociation.

Inner beings have a mind of their own, complete with their own unique values and memories. They co-evolve with each other and the world that they inhabit. Carl Jung referred to them as complexes, both personality fragments as well as enduring characters. They have their own agendas and energy. When stimulated they burst into our consciousness. They have the ability to influence outer objects and people from beneath, Explained as charisma, synchronicity, poltergeists and other curious forces. They can travel outside of our minds as characters in books and movies and into the inner landscapes of others. These characters are gained throughout life. Families and cultures share them with each other and they are passed on to the young. Some have great influence over our thoughts and actions, others very little. Their relationships with each other are mirrors of our outer world. Their society is patterned by our beliefs and fears. A man whose life is dictated by an oppressive hierarchy of bosses and kings, has an inner world peopled with slaves and masters, living in slums and palaces. A man raised in a more egalitarian culture may let the inhabitants of his mind run free.

By being honored and embraced, beings are empowered. Paradoxically, they also gather strength by being feared and hated. Repressed into the shadows they slowly simmer until they smash their prisons and explode to the surface. Here they struggle for power with the forces that have shoved them down and deformed them. Inner beings have an archetypal core which is expressed when their part in the mythology arise. Driven by deep-rooted instincts, they emerge when they hear their cue.

The most ancient archetypes play the greatest parts in our personality. These are the gods, the avatars, the angels and demons who quietly or explosively shape our world from within. Never expressed in their pure state, they form the spiritual core of beings and objects. They clothe themselves in different bodies to keep up with the latest styles and beliefs, but inside they stay true to themselves. As language can never do justice to the gods, they are revealed to us in symbolic forms. Even as the names of the gods change and symbols slip deeper into the collective mind, the strength of these primeval beings arise when pathways and plotlines conjure them up to fulfill their destinies.

I inherited the very shell of somebody's' life and used my imagination to breathe him into being. I then began to eat the same foods as he, bought the same car, read the same books and sought out similar friends and lovers. I nursed him inside if me so that his spirit became the most influential being in my inner landscapes and dreams. He then recreated me in his image. What began as self-conscious posing became natural and flowing. I needed him because my substance was weak, so he burned it up then lived on as pure fire. We then walked in both worlds. I existing as a character in his dreams, a prophet warning him of things to come. He, a detective and philosopher in my dreams informing me of how we ever got to this point.

Our inner world is fed through the outer. In dreams we observe a rich parody of our lives in the solid phase. We sleep as scuba divers plunging into the Great Ocean, leaving that tiny nucleus we consider to be reality. Within the nucleus exists the Grand Project, a master creation given continuity and structure. In this strange realm time actually seems to exist and things are solid. A mathematical framework serves to make this world logically consistent. It is the central dream that all others revolve around. Outside the nucleus is the protoplasm of anything, where minds travel without friction to other worlds. It is the background of being that we are taught not to see, the darkness that enlightens the stars. It is the quantum realm, where the laws of physics are mocked and diminished. Here time is liquid and unembodied and matter reveals it's hidden life-force. Visions and dreams aren't chained to the laws of perspective. The spirit is unencumbered by gravity. All the thoughts thought are patterns of plankton, seeking associations and molding ideas in the cytoplasmic gray matter of the Great Mind. We are the complexes of this larger being, enduring pixels of light with memory. We shape the inner life of this entity as it dreams us into existence. We hardly realize the symbolism of our own actions as our thoughts initiate us onto pathways and plotlines. We become the mythological characters of the gods, as much as they are to us.

Is it worth it to become a god? Throughout history, religious and political leaders have been inflated to the level of the gods by the adoration of followers. Our cult of celebrity chooses a select few mortal humans to be raised up as symbols of the zeitgeist of each generation. We willingly hand over our power to the chosen, turning a person into a solar system. Through our worship, we craft an inner deity that has great power in shaping our personas. We begin to mimic their mannerisms and clothing, as their star rises within us. The movies and news clips that they inhabit are our created dreams.

There exists a paradox in the formation of gods. Did they create us, or did we create them, or do they rise simultaneously as the dream defines the dreamer. There is a major drawback to stellar evolution. As someone's life is increasingly made public, they lose a degree of free will, as the audience begins to script their lives for them. The masses create their gods in the world of their thoughts. The stars respond like marionettes, impaled by the strings of their projected image. The great power that a god is gifted with, is tempered by their inability to control it. The public lives in the inner spaces of the mind of a god, just as the god exists within the people. They become our angels and demons that interact within the boiling sea of our personality. We are their subjects, but also their creators. It is our imaginations that have summoned them into being.

Whether or not a god existed in a physical form is irrelevant to its power. It is unnecessary to have a material body to be a god. In their truest state, gods are ideas formed in the deepest regions of ourselves, where personal imagery gives way to collective mythology. Apotheosis occurs within a person when they channel an ancient archetype that coincides with the spirit of the place and times in which they live. The man or women that is thus possessed, lends their image to the risen deity. Following the death of the body the mass mind distills their spirit by cleansing it of the dross and dregs of daily living and ironing out inconsistencies of character.

No actor can play a role that is not within them. The best actors are able to bring an inner being cleanly to the surface, as they split that aspect of themselves away. Many actors become possessed by the characters that they become, as does the audience. The movies that they perform are their dreams. Each part, a further realization of who they are. When an actor is typecast, their personality becomes stagnant in a tangled web of expectations. It is the expectations that create the star out of a person, distilling an essence and neglecting the complex whole of their personality.

The god outlives the being that birthed them into the outer world. They live on and evolve in the minds of fans and followers. Fictional characters are no different. Springing from the mind of a storyteller, characters enter into the public consciousness as living beings. Hamlet, Bugs Bunny and Moses are evolving life forms, changing over time as they enter the minds of new generations.

Sometimes gods just fade away with the dying of a culture. Some are later resurrected by the cycles of history. The most powerful deities are those with the greatest number of highly devoted followers. It is the act of worship that raises them up. Our beliefs about the world coincide with the ascendancy of a god that is empowered by that paradigm. Even an atheist helps to empower a god. The idea of a soulless universe empowers the archetype of the Machine. The god of the Bible also has great strength in a world built of conquest and empire, lords and bosses, dictators and kings, as does Jesus, who may represent the oppressed rising up to their oppressors, by finding the light within.

Often, gods are used as symbols by those in power in an effort to manipulate people. Their evolution becomes consciously directed and their spirit becomes domesticated. Coca Cola gave Santa Claus a red and white suit and helped change him from a generous spirit to a symbol of consumerist greed. The Romans took the free spirit Jesus and turned him into what he most opposed, an emperor.

Idolatry is unavoidable. Even if we were raised without celebrities and politicians we would idolize relatives, ancestors and animals. To worship is to live. It is to point the light of consciousness outward at someone or something and raise it up from within. There is a great danger when our heroes are chosen for us, as we can then be manipulated down pathways, not of our choosing. Instead of spontaneously erupting into our thoughts, gods are now engineered with specific goals in mind. They are often used as a diversion, to draw us away from what is important. They are like magicians who actually destroy the idea of magic, because their powers are staged.

The most ancient gods are the ones closest to the source. Deities live within each other, just as cells form the body. Every layer of reality lies nested in another. The chain of being forms a loop, where the largest and smallest substance is the same. Plants, stones and animals far predate humanity. Their spirits are the ancestors of our gods, the most ancient of which are most worthy of our reverence as they touch closely upon the source of being. Through them we can travel towards the roots of what we are. Nameless and faceless, they are the wind that animates lesser gods.

The Other cannot exist. All that is must be One. Objectivity and separation are symptoms of a prescribed myopia. The Other is our Self unrecognized. Somewhere in our travels we have forgotten who we are and have walled off our highest being. We have undertaken a massive construction project in our minds to compartmentalize the world. Our language is used like a sledgehammer, breaking everything into little pebbles. We've created a giant shadow where we hide our true identity. In the shadow we also conceal that which is secretly controlling us. We have built prisons to enclose parts of our selves that we refuse to acknowledge. We have ghettos to keep the dispossessed. We have factory farm hells where we secretly torture the free spirits into submission. We have millions of third world slaves building our egos with their sweat. In the shadow we imprison all that we despise or fear and label them Other. We point our fingers outward due to our failure to look in. We reconstruct our external world to mirror our inner. Genocide is the engineering of our soul. We make war on our planet forgetting that we're attacking our own body. We have cops and armies to subjugate those that try to break loose. We make it illegal to walk around undisguised. We massacre the free running creatures by the millions. We put cameras on every corner feeling that the enemy is ever closing in. We abduct the uncontrollables and bolt them in stone. We refuse to own up to our own actions, which have given rise to legions of dark phantoms seeking either revenge or redemption.

Many beings grow warped in the shadow, building strength in their rage towards their oppressors. Saboteurs, revolutionaries and terrorists explode to the surface where they attack the system that enslaved them. Civil war is the dark night of the soul, where caged beings breach their enclosure. If nothing is learned from the fight the system rebuilds itself, except the oppressed becomes the oppressor. The cycle repeats itself in an endless series of revolutions, each side believing that the other is the enemy that deserves to be eliminated.

We can't project a shadow onto others, as projection implies putting something outside of ourselves. It is instead a resonance, a participation of shared aspects. That which we most fear about others resonates with the beings of the darkness within. That which we have disinherited, enters into our lives as a stranger, one day revealing itself as an ambassador from our cast shadow.

In higher states of mind all dualities are erased and all fractures are healed. The wind becomes an extension of our thoughts, weather and mood inseparable. Trees and animals lose their indifference towards us. We see things most clearly from the corners of our eyes. We can know others as we feel them from within. This state is the birthplace of compassion and the definition of love. It is where we find out who we are.

We are discouraged from connecting with our source. The knowledge of who we actually are is prohibited. Mind doctors label our higher states as childlike, primitive and subjective, the projection of our thoughts into a mindless universe. It is categorized as schizophrenia, a sickness that needs to be treated with drugs, surgery or high voltage.

Could all of the native peoples of earth be insane? Is it really possible that all of the thought of pre-enlightenment man is delusional? How could it be that people who believed that stones and trees speak possibly survive? For not only did they survive they often thrived.

A world-view that is deluded and deranged and forged of partial truths would be a destructive path to follow. Cultures living a lie would soon die off along with the land they inhabit. Our ancestors survived millions of years by following their visions, their minds interwoven with the living landscape. They live on within us though we scarcely listen. It is not they who are deluded. Our way of thought has devastated our world, yet we continue on this path. How have we been tricked to believe the ultimate lie? Why do the world's religions remain idle as God's sacred creation is eviscerated? Some of us believe that heaven awaits us, so to hell with the earth. Some of us see the earth as a resource to be exploited, so we blind ourselves to the carnage. But for most of us, we just refuse to think about it. We keep busy at work, never knowing what we're working on. Missionaries of nihilism guide us through school, teaching us about the universal machine that indifferently hones us into properly shaped sprockets. Restless, but drained of feeling and curiosity we become patsy slaves to those that control us.

We can sense that something isn't right but we can't place it. In exchange for all we have lost we gain an insatiable appetite, consuming all that we can to fill the empty part of our souls. We are told that our dreams are meaningless then we are told what to desire. Heavenly bliss and wholeness await us if we walk the pathway that has been engineered for us. It is called the American Dream. Now trademarked and distributed via satellite, so that all earthlings can now fantasize of great mansions, control over others and the adoration of millions. No matter how many of those that have made it die in their excess and emptiness, we still buy into it. The characters of this dream are revealed to us not in our sleep, but through our televisions. Their faces are plastered on buses and buildings sending their message of transcendence through consumption. Instead of listening to inner teachers to guide us, our goals and beliefs have been hijacked by specialists in mind control. The science of propaganda once used to instill beliefs in an ideology, has been perfected into the respectable fields of public relations and advertising. Masterfully manipulating sacred symbols and consciously scripting mythologies, advertising subverts and conceals our inner life and replaces it with slogans, jingles and salesmen. Selling products is only part of what commercials accomplish, they also create a product, the consumer. This is a person who is always unsatisfied, restless, bored and insecure, forever seeking a new potion to alleviate their anxiety. A spiritual vacuum is necessary to convert the masses to the religion of consumerism. This manufactured alienation subjects people to control by totalitarian governments we call corporations. Each new generation is exposed to increased levels of mind-control that is often thought of as entertainment.

"Not to know ones true identity is to be a mad disensouled thing – a golem, and indeed this image, sickengly Orwellian, applies to the mass of human beings now living in high tech industrial democracies. Their authenticity lies in their ability to obey and follow mass style changes that are conveyed through the media. Immersed in junk food, trash media, and crypto-fascist politics, they are condemned to toxic lives of low awareness, sedated by the prescripted T.V. fix they are a living dead lost to all but the act of consumption."

Terence Mckenna
Food of the Gods

To lose touch with our source is to be unable to feel that other beings are a living part of our selves. Even under the solid façade of objects lurks a vitality that we can only know by feeling it from within. Through our imaginations we can feel the living aspects of an object. By doing this we slightly change the object by bringing forward what we are able to see. This is similar to a sculptor bringing out the hidden form or feeling in a piece of wood or stone. Many native peoples similarly view objects using their imaginations. Not projecting their thoughts onto lifeless things, but resonating parts of their mind with qualities inherent in an object. Perception and observation are always a creative process. Preconceptions and prejudice limit our ability to know something to those qualities that are expected. Scientific experiments are limited by the narrow beliefs of the system, forming a deadened world to match expectations. Even as quantum physics reveals an observer created reality, other sciences still insist that their experiments are objective. Science is masterful at discovering outer qualities such as hardness, structure or mass. It can even find out such abstract qualities as the amount of energy stored in an object. Hundreds of years ago only a mystic might have known that a speck of sand contains enough energy to level a city. Now it is so commonplace that we cease to wonder how matter holds such power within its seemingly stable solidity. Science fails to investigate the immeasurable inner aspects of matter, denying it's living aspects in the process. Life and mind cannot be quantified therefore they are disregarded.

Objects exist as the physical manifestation of the line of thought that brought them into being and are imbued with the qualities of that which created them. Man -made objects are no exception. But as they are increasingly mass produced, depersonalized and disposable they are rendered as soulless as the conveyor belt that they've rolled off of. The unique and beautiful objects that mankind had once possessed have been replaced by sterile, utilitarian products manufactured by machine. Even so, any object can be imparted with soul and character over time by its possessor. A pair of shoes holds a memory of all the miles walked, a favorite hat gains the character of the wearer each time it is put on, a car responds to care by becoming more alive, a house resonates with all of the joyous events and tragedies that have occurred within. The greater the emotional attachment or significance imparted to something the more soul it reflects. Every object is like Pinocchio responding to the love of those around it and coming to life. An object is a possession because it possesses, it enters into our inner life and interacts in our dreams. An object is also possessed by its owner, as her spirit enters into it and helps guide it's evolution.

Social pressures push us to discard our old clothes and cars soon after they are acquired. Objects are designed to self destruct or else the technology behind them made obsolete as soon as they are bought. Buildings of glass and steel alienate the living. Teflon and scotchgard repel character. Styrofoam stubbornly refuses to integrate into the landscape. Modernism is a war against the soul of objects, an overreaction to our haunted past.

Nature moves all things toward a unique expression of them selves, while simultaneously integrating them into their surroundings. Even the deliberate objectivity of modern artifacts becomes softened by the subtle fingers of time. Uniformed paint becomes dappled, plastics fade and crack, smooth metals rust and pit, straight lines and hard angles curve and dip. Eventually all things become a coherent part of the landscape.

Naturally formed objects have had eons to evolve diverse patterns of growth. A log fallen in the woods becomes a canvas for natures' artistry. Wind, fire and rain, along with, mushrooms, bacteria and animals work together with the intrinsic character of the wood to sculpt complex and beautiful forms. A beings location heavily sways its evolutionary pathway. Wood, having fallen into running water is transformed into a materialization of the waves and eddies of the flows it inhabits.

Artists have always wondered about the images that appear in natural objects. Michelangelo claimed that he liberated figures hidden in a block of marble. Endless stories are told in a patch of mildewed wall. The forms of rocks and trees give clues to their inner life. Paracelsus tried to make a science of knowing the "inward virtues" of natural objects through the Doctrine of Signatures. The idea behind it was that the creative powers leave their characteristic marks or signature on all forms that they shape. The Chinese used a similar system. For example, a ginseng root, which grows in the form of the human body can be used as a general tonic. Walnuts, which resemble a brain inside a skull can be a treatment for mental prolems. The problem with any system that becomes indoctrinated is its lack of flexibility. Reading into the mystery will always remain an individual and subjective process in an evolving world. Secrets revealed in one age become meaningless in another. There is a paradox of the natural world, that clues are left everywhere for us to discover, yet natures' true essence stays one step ahead of our intellect. The holy Trickster erupts into being every time that we decide that we really know.

Wild animals live close to a state of grace deeply connected to the core of reality, where boundaries are transcended and all beings unite. Animals often reveal themselves in dreams before making their appearance in our outer realm. Their ancient archetypes appear and evolve in our inner worlds. They are the gods and goddesses of the ancients and the alternate forms of anthropic deities. When their essence is strong within us they are our totems, our angels and sometimes our demons. The pantheon of the animists populate our inner ecosystems as well as our earthly heaven. Here they live their lives as individuals, each sculpted by their own unique history. Their eyes perceive the beauty of creation from every possible angle. Their colorations and graceful forms are glories for others to receive. Their cycles and rhythms are rituals, replaying the drama conjured up in the dreamtime of their genesis. Because they are most like us, they offer a reflection of our psychic makeup and open portals of insight into deeper regions. Night animals inhabit regions of mind where the light of our consciousness struggles to penetrate. They can be messengers from the darkness where we've hidden the secrets of who we are and why we are here.

There are two different but complementary ways to be. All things are both a part and also the whole. Evolution strives to achieve a balance between the two states. The ideal is to promote free will and individuality while protecting the integrity of the larger body. It is both the utopian goal of civilization and the Eden of our often forgotten past.

Arthur Koestler described the two tendencies of being as self-assertive and integrative. The self assertive state is one of separation, ego, competition and greed. It is the world of solid, objective, matter. The Maya of the Hindus and the reality described by Newton and Darwin. The integrative state connects the parts as a whole, it is the transcendence of self. It is the myth or the archetype that draws things together. It is the source of love, altruism and cooperation that leads to health and sanity. Integrative emotions include empathy, wonder and rapture. A dynamic equilibrium grows between parts and whole. Under conditions of stress the affected part may break away creating an imbalance within the larger being, causing illness or insanity. A diseased state exists until homeostasis is regained. Scientists that study the chaotic nature of dissipative structures describe the process as a bifurcation caused by an instability in the system. This also describes the formation of new species as a single, stable field bifurcates into two. The new species must then integrate into the greater field or face extinction. It is the branching process that patterns our thoughts and our evolution.

Energy also exists in these two states. It is the duality of particle and wave, body and mind. Particles are the self-assertive aspect of energy, giving durability to vibrating patterns. Waves integrate energy into forms shaped by morphic fields. When matter dissolves into waves it loses its qualities of separation and vibrates in an indivisible oneness. The state that matter exists in depends on the consciousness of the observer. If we approach matter from the perspective of an objective outsider, our consciousness is a narrow beam of light separating an object from its context. This gives us the impression that an object has a solid independent reality outside of Mind. In a higher, contemplative state, consciousness is like a floodlight that reveals matter as a network of relationships. The observer and the observed are inseparable and dependant on each other for existence. The backdrop and foreground are revealed to be indivisible facets of the same picture.

Like matter, mind can also exist as seemingly independent units and also combine into larger entities. The self-assertive aspect of mind is ego. It is what gives us the illusion of separation and material reality. Ego is the protective membrane that protects an entity from dissolution. It is the transient and permeable border between beings. An amoeba under a microscope is its own being, it is self-enclosed and mobile. From a distance the amoeba is also the pond that it inhabits. It is the pond that gives it a context, that shapes it's evolution and gives it a reason to be.

Not all integrative trends are positive and lead to harmony. Disease is a self-assertive force, as bacteria or cancer compromise the bodies' wholeness. But within the disease is an integrative force that unites the germ colony as a single being. Similarly, a person may be well integrated into society, but if that culture isn't in balance with the earth then that individuals actions becomes destructive. Natives living in balance with the earth are considered to be a disease to those living in an earth destroying society.

Some forces defy simple classifications. Communication within the body integrates tissues with hormones and electrical impulses, but it is also self-assertive as it is a signal between separate entities. Language can unite people together, as myths and ideas are shared, but it also stresses the separation between speaker and audience. In conversation speech is a wave that passes between people, whereas when people sing together the wave unites and all separation no longer exists.

In language, nouns give the impression that there exists independent beings or objects. After every noun we could imagine a symbol $\circledcirc$(∞), this represents the All. Implicit within everything is all that there is. In math we can place the All symbol after every number to remind us that numbers are an abstraction, and don't have independent existence. $8^{(\infty)} + 4^{(\infty)} = 12^{(\infty)}$

At the center of every individual is the All. Every place is the center. Intuitively, we can feel this. When we drive down the road we seem to stand still and things in the distance come to us. Everything seems to revolve around us. The earth seems to be center of all things, even thou we revolve around the sun. Or do we? Is it not possible that the sun and the rest of the universe revolve around the earth. Mathematically, it is possible to prove that we stand still while everything else moves around us. Pick your arms up and spin around. Is it you spinning, or the room around you?

If the center is also the whole, this would imply that the largest and smallest numbers are the same. Instead of the number line stretching out in opposite directions we would get a closed loop that joins inner and outer, large and small in a multi-dimensional mobious strip.

The idea that the whole exists within the part can silence many a philosophical debate. The hierarchy of being doubles back on itself, enclosing the infinite within the mortal. Notions of death and life, time and eternity, mortals and gods, free will and fate are unified under a central paradox. The idea is revealed to us in a hologram, where each part contains the whole. It is the DNA within a cell that can create a world. It is the indivisible God that resides within fragmented man.

Our culture emphasizes the ego and neglects the whole. We are left enclosed in philosophies of existential loneliness, obsessive reductionism, Darwinian competition and narcissistic consumerism. Each representing a hurdle to the natural growth cycle of the Self. When we are conceived into this world our minds are an undifferentiated extension of the universe. Our neurons are a network of pathways reaching out in all directions, ready to learn and adjust to the particulars of our life. As we grow up our pathways are narrowed to match our experience. Most of reality is filtered out as we are taught to focus on a narrow band. The flowing imagery and wonder of childhood gives way to the forced march of the adult world. Our personal reality encapsulates itself in a protective shell, allowing only certain things in. The formation of an ego is an important part of development, the expression of an independent self. We are taught that it is the last stage of development, neglecting the circularity of life. To continue to strengthen the ego is to seal off from the source, leaving just an empty persona where someone genuine once stood.

Following the formation of the ego, the next stage is to explore what unites us. A search for truth leads us full circle to the unity we rejoin when we die.

The ideal of the wisdom of elders has given way to the uselessness of the old. Atrophied nursing home inmates are symbols of our loss of life's' continuity. Death is regarded as a failure in our absurd war against the darkness. When the Ego is mistaken for the All, dying is regarded as an act of tragic finality rather then the fulfillment of life's cyclical promise.

A baby's face reflects the wisdom of the aged. Assuring us that life isn't a line from past to future, but a spiral path bringing us home to where we started and hopefully wiser for the journey. Our ego is the part of us furthest removed from our true self. It is able to venture into our collective mindscape and believe that it is something other then itself. For thousands of years now, the deconstruction of the divine and the desacralization of nature have left the ego supreme. The reigning gods of separation teach the values of objectivity, competition and selfishness. As if to compensate for this imbalance, Eastern religions have swung so far the other way that the ego is obliterated. The danger in this is neglecting the earth, the body and the individual. The fact that we have materialized here on earth has meaning, it is a major assumption to underestimate the importance of coming into being. If all was to forever exist as an undifferentiated unity, evolution would cease and the beautiful and sometimes tragic story of life couldn't be told.

Many states have failed because the freedom of the individual is forfeited to the power of the collective. If the collective were the earth instead of a government, a better balance between individual liberty and planetary coherence would be achieved. Nature evolves unity from individuality not conformity. A government of even the most perfectly united comrades would be a cancer to the planet if unintegrated into the whole. Each nation represents the formation of a meta-ego encapsulated in its imagined boundaries. The more that a people are separated from the land they inhabit the greater the dissociation of their country from the Earth mind. The dream of uniting humanity into a global order of mankind is a destructive nightmare for the planet. For the coherence of nature is formed out of a lack of central authority. We are given the terrible choice between remaining as warring nation states or uniting to form a world government. We are dividing humanity into ultra-conservative fundamentalist chiefdoms and malignant corporate controlled mass culture, each option is corrupting in its own way. Global capitalism is especially devastating to the ecology of the earth, as high-tech humanity has come together to cannibalize the planet for short-term gain. The diverse languages and cultures that have grown out of the land are being replaced by an engineered monoculture that is making every place the same. The long-term stability of our world is dependant on it's diversity which is rapidly being destroyed.

We are spurred onwards by the myth of progress. It is the idea that if we take charge of our evolution, we can make the world a better place. We neglect to notice that every gain in one area, creates a loss somewhere else. Attempting to direct our evolution creates growing cycles of ever -greater imbalance. Control over one aspect of our life, necessitates only further domination. We are told to watch what we eat, control our tempers, mow our lawns, to use our wills to subjugate our mind, body and world. Implied in these actions is a mistrust in natural order and wisdom. Every part of our lives has been subject to the science of management. We consent to this intrusion on our nature out of an implanted fear that things will get OUT OF CONTROL. This fear has created a gigantic infrastructure of coercion and incarceration, a system of control so complex that it requires twenty years of education to learn how to correctly fasten the chains.

The greatest irony of all, the biggest joke played on humanity is that all of our work has made things worse. This project of ours, we call civilization has been a failure because we live in a universe that functions best when it is free. This isn't a philosophy easily translated into language. It is the mind of wilderness. It is written in the stars and etched into the bark of trees.

Our instilled fear of disorder and entropy has driven us to believe that we need to be governed. The paradoxical nature of reality has now led us to the realization of our fear.

Our dreams of conquest are born of dissociation. They are an absurdity uttered by a manufactured persona. There is nothing to gain, because we already are everything. There is no progress, because at our center is perfection. That perfection can only be realized into our world if we release the stranglehold we have placed on our souls. Love becomes deadly when it is possessive, fragile flowers wilt in the tight grips of our hands, animals lose their shine when locked in a cage, our blue and green body withers at the implementation of our plans.

Development

A bridge with no river
A tall façade with no building
A sprinkler on a plastic lawn
An escalator to nowhere
A highway to the places the highway destroyed
An image on a TV of a TV showing another TV on which there is yet another TV

Eduardo Galeano
Upside Down

You don't exist! If I am consistent with the philosophy espoused in these pages, then You are Me. To acknowledge your existence is to be seen talking with myself, and you know how that kind of behavior is frowned upon by society.

A paradox of speed is rapidly encroaching on our lives. As we have come to fear the passage of time, we've exponentially sped up our lives to supposedly give us more of this precious resource. We fill our habitats with "time-saving devices" and schedule our days to fill every last minute. We use computers that calculate at astronomical speeds. We build super highways and jet planes to bring everything near. What we don't realize is life's ever-present fondness for irony. All of our efforts to cram our days with high-speed transactions are negated by the resulting quickening of time. Believing that we are arriving at the promised land at an ever faster pace, we can barely stop and look around. If we could, we would come to the conclusion that we are only chasing our own tails in high velocity revolutions. Heaven recedes as we race towards it. We arrive when we stop running only to realize that we are back where we started from.

We once dreamt that we could travel to rare and exotic places with little effort or time. We built magical machines to shrink our globe so that all of the worlds' mysteries were at our doorsteps. The realization of our dream was far from what we've expected. The faraway places that were once so distinct had become twisted facsimiles of home. The enchanted lands that once had been became gilded attractions in a global theme park. They were parodies of what they once were, cultures that were bought, sold then reenacted.

High speed travel blurs the landscape until all becomes indistinct and homogenous. The adventure of the journey itself has becomes a forgotten remnant of a time when it gave us the greatest reason to be.

The only way that One could set it's mind into motion was to imagine that Zero and Two actually existed. Zero was a true phantom, its reality logically disproven by the One; "If I Am, then Nothing is Not". Two was just as illusory; "All that Is will Always be One". One when multiplied, contrary to mathematics will still be One. Even when infinitely divided, the fractured numbers are all of one piece. One first envisioned then blinded itself to create the other, as the drama of being is dependent on the many forgetting that they are the One.

Our memories are flawed and therefore we are blessed. We cannot truly be sure of anything and therefore we are in awe of living. The visions that we construct in piecing the past are the mythologies that we use to build the future. Exaggerations and understatements become lies, become fables, become truth. Our biased interpretations of past events are attracted toward the mythic. Wishes and fears are reflected in what we choose to acknowledge, and what we can't help but forget. Contrived history shapes the mythology of the coming age.

Life proceeds by forgetting itself. Imagine creating a world where you oversee all of evolution, a universe in which every detail is produced with a full awareness of its presence and destiny. All freedom would be erased and all limits reached in an instant, as the first instant would imply all others. Evolution would be pointless. The universe would be a stillborn artifact of a finite and one-dimensional creator.

It is the unending revelation of mystery that liberates existence. It is the unknown that makes life interesting. Only in forgetting oneself may an omnipotent deity attempt to fill the boundaries of what may be.

The unity of our world relies on the diversity and individuality of its parts. Rationally this statement registers as nonsense. How can freedom and anarchy achieve the greatest harmony? The answer hides deep within the relationship between the One and the Many and deep within the soil of a wilderness forest.

We misunderstand the deeper nature of reality when our overriding analogy of the universe is that of the machine. We engineer our machines to exacting precision, standardizing the shape of each gear. We are led to believe that the slightest irregularity within each piece is the future ruin of structural integrity. Our devices run for a few years and unable to repair themselves they wear down and break. In comparison the Earth grows from within, a strange profusion of unique, whimsical and inefficient characters. Somehow in its ever-changing grandeur and obsessive complexity the Earth is self-renewing. Our denial of this paradox and our stubborn insistence on reason has lead societal engineers to insist that unity be achieved through monotony and standardization. All around us we see evidence of the mass engineering of our world. Instead of allowing the beings of Earth to grow freely we have assembly lines, monoculture industrial farming, standardized testing, genetic cloning, mass media, global franchises, multinational monopolies and world government. We pretend that our culture values the individual, yet our personas are as mass produced as our products. Goose stepping soldiers and monoculture tree farms are iconic of our times. The fascist obsession with synchronization and order has survived its bad P.R of the holocaust in its new disguise as corporate efficiency and streamlining.

All of the popular political ideologies insist on our being controlled. Many well intentioned progressives have come to believe that the future must be managed and ruled. Socialists and Internationalists pin their hopes for the future on having even more centralized forms of authority. Libertarians and Neo-conservatives claim to be in favor of less government, seldom realizing that the corporations that they support are forms of totalitarian regime.

Much of the ruling class of western countries busy themselves with details of administration and supervision. Their skill is in mass control. Raised on university beliefs, lawyers, biologists and engineers carve niches for themselves in the field of planetary management. Who amongst them is prepared to deny their expertise and claim that the Earth and humanity evolve best without professional intervention.

By allowing all creatures to be that which they are, the earth retains its coherence. Apart from the controlled world of man, we don't see one species enslaving another. Individuals eat each other but don't commit the infinitely worse offense of malevolently possessing the soul of their prey. The leopard feasts on the antelope, but is not its master. The frog eats a spider, but without genocidal passions. A deep moral tendency prevents one life form from keeping another in endless bondage and existential suffering. Our own feelings of compassion or guilt towards those that are held down are an extension of this unwritten code. Striving for Universal Freedom and Justice are ideals exaggerated in civilized humanity, specifically because of our cultural failures to deliver them. Superman is born of a broken world. Our treatment of fellow beings swings from extreme acts of kindness to the worst depravity imaginable, as our world desperately attempts to right itself. Our lives would be deplorable without those willing to give of themselves to help another.

In accordance with Darwinian orthodoxy, kindness is non-existent. It is actually a brilliant form of deception that has evolved to increase the virulence of our nucleic acids. The big three religions trace the Good to the God, evil is rooted in a severed entity. However among the meditations of mystics the devil is in the duality.

Morality need not be based on some supernatural lawgiver and judge, but in the unity of all life. We can deceive our minds into believing that there is something to gain, some profit to be made in confining others and milking them of their life-force. In a larger sense the Other is always within the Self and consequently, the rabid vampire is really biting his own artery. To see an absolute division between one and another is to be blind to the Love that unites them. It leads to a failure of the imagination to realize that nature follows the mythological Golden Rule. In wild things, one comes across an overwhelming acceptance of death and life, a contentment rarely experienced in those that are governed.

There is in nature an equalizing force. Intuitively we sense a striving for balance. As bipeds our bodies know to keep our spines aligned with the axis of the world. As accountants and gamblers we realize that numbers have a need to cancel each other out. As scientists we know that every force is countered by an equal and opposite force. As emotional beings we perceive the narrative and counter narrative of karma and the circularity of life that it implies. Yet as statisticians, we might flip a coin 200 times and always get heads and still insist that the odds of another head is 50:50. Intuitively we might think that another head is likely if the trend is to continue. But for the final count to be even the we might believe that the odds must skew towards tails. For this to be true the equalizing force must be intelligent, something few mathematicians would acknowledge. Even still, the odds eventually even out. Somehow the books must be cooked to include in this world both precognition and profitable casinos. The equalizing force is like the dark matter of numbers, spurring many minds towards wrong choices to counter a single psychic sure thing. Good luck and bad annihilate each other like matter and antimatter in order to lend a modicum of stability to our waking life.

From Marxism to Taoism every culture acknowledges the struggle of opposites, gains on one side equals deficiencies on the other. To see the opposite as complementary is to know that nothing can truly be gained, but only changed in form. To know this is to puzzle at the ideas of progress through technology or victory through war. In the end as in the beginning All is made Equal.

On further reflection the equalizing force dissolves in a deep asymmetry. The indisputable fact that something does exist reveals an eternal bias towards being. An infinite number of non-existent worlds could never counter this moment. Even a mirror world composed of anti-matter complete with backward time could never balance what could only be deemed as a miracle... WE ARE HERE!

Before the arrival of man, North American landscapes rumbled with the presence of large furry beasts. Coinciding with our appearance in new lands was mass slaughter. Here was home to elephants, giant ground sloth, glyptodonts, giant bears, saber tooth cats, cheetahs, massive beavers, horses, tapirs, camels, llamas, ox and yak. By 10,000 BC these were among 57 species of large mammals that were wiped out. Extinctions in South America included entire orders of creatures, and 32 genera of small mammals as well as horses, hippos and mastodons. In Australia all the big mammals were eliminated except kangaroos. Destroyed were the marsupial equivalents of many placental mammals, including lions, ground sloth and a 2000lb marsupial rhino. The marsupial wolf survived into the 1930's. Giant birds, tortoises and snakes also disappeared, coinciding with the landing of strange slender primates. In Europe, Asia and Africa the extinctions were drawn out over time, matching the flow of people from Africa. Each conquering wave brought new technologies, customs and a lack of attachment to the new land.

In Africa, many large mammals survived as they co-evolved with people. Even still, the death toll of African mammals has been huge, including many monkeys, apes, and hominids. The carnage is ongoing and has speeded up in recent years. In all likelihood, within 20 years the wild and free elephants, rhinos, cheetahs, giraffes, and zebras will only survive in confinement, if at all. Their remaining habitat has become a safariland lunapark.

The greatest evidence that humans have destroyed the large mammals of the world is that it is still going on. Habitat destruction and outright slaughter are more widespread then ever. Animals once too numerous to count have succumbed quickly to human arrogance and insanity. Buffalo, which numbered in the millions were reduced to a few, as part of a genocidal campaign to starve the native people. The passenger pigeon which once filled the sky with flocks of a billion birds, were shot out of existence by crazed hunters fueled by blood lust.

The heart- breaking story of destruction caused by invading Europeans on the Americas included not just the Indians but all of the beings that were considered most sacred. Bears, wolves, and birds of prey carried bounties on their heads. In South America, insatiable greed razes the forests of the tropics. Destroying thousands of species that inhabit this diverse dreamscape, that is the very soul of this planet.

Sea turtles that have flown these oceans for hundreds of millions of years are discarded by shrimp boats like yesterdays trash. Beaches that were once filled with bright eyed hatchlings, lie sterile and empty, or else covered with concrete and neon. Our oceans, once vibrant and teeming with life, suffer a slow death as a convenient dumping ground for sewage, industrial and nuclear waste. Factory trawlers with thirty mile nets, filter the ocean of life and grind it into a slurry. Mangroves that haven't been replaced by seawalls are ripped out to make way for horribly polluting fish farms. Salmon, which once spawned by the millions are blocked by dams that not only destroy entire ecosystems, but the spirit of the river itself. Coral reefs blown apart by dynamite and poisoned by cyanide fishing, marshes filled in, prairies dug up, mountains bulldozed, forests set ablaze: the list of transgressions against our fellow earth beings are to numerous to ever complete. Everywhere we see signs of sickness, dying trees, soil erosion, desertification, acid rain, viral epidemics, killer algal blooms, eutrophication, coral bleaching and climate instability. Our actions are responsible for one of the most destructive eras our world has ever seen.

The process of invasion is a necessary part of thought and evolution. Every new idea upsets the balance of mind until it integrates into the whole. Despite the destruction left behind by the colonization of new land, many peoples eventually integrated into the ground they lived on and became a part of it. More then just a place to exploit, the land became an extension of themselves, in tune with their thoughts and responding to their needs. A connection was formed, erasing the dualism between culture and nature. The desire for expansion and material progress was exchanged for natural growth rhythms and artistic expression. Technological evolution was replaced by spiritual growth. Many cultures around the world have thrived with only the minimum of tools. When in health, the land provides its inhabitants with all they that need to live. A deep wisdom evolves over time based on communion with the land and it's beings, making the ego permeable to the larger mind. As evolution moves towards harmony the need for work disappears, and people gain the beauty and intensity only found in those that are free.

Our journey towards grace was not to last. In many places, waves of conquesting peoples wiped out the wild humans with new weapons and attacked the landscape with new tools. Each new development carried the power to damage the land and its inhabitants, as no immunity had evolved to deal with the conquerors technology and behavior. Ecosystems were unable to keep pace with technological change and eventually collapsed. When the land was too degraded to support them, the invaders moved on to the next place to ruin. Forests and peoples fell due to forces that were unprecedented. To survive, tribes had to take up arms to defend themselves, and in so doing became like their marauders. The world was already at war by the time of the new arrivals. It's a miracle that some peaceful peoples such as Hopi or Bushman survived to this age, their remoteness and marginal land being their ally.

Waves of empire began to divide the people of earth into the conquerors and the conquered, victims and victimizers, first world and third. The free humans were enslaved and domesticated. As they disappeared, empires set the blades of conquest against each other. People walled themselves in against relentless assaults. Arms races spawned high technology. Empires fought world wars against each other each trying to possess the world, creating a self-feeding model of conquest. The earth's energy was diverted to the war machine. New technologies were created to impose ever greater levels of control, as frontiers of further conquest were discovered in the genome and the brain.

NATURE ← Each contains the other→	**CIVILIZATION**
(Balanced evolution, Wilderness,)	(Man-made, Enforced Imbalance)
-civilization freed from control	-nature supressed
-no one takes more then it needs	-rich elite, impoverished masses
-unruled and free	-authoritarian, domesticated
-moves towards greater diversity	-moves towards monopoly
-evolves as an artistic masterpiece	-mostly utilitarian
-infinite depth and mystery	-often shallow, superficial
-cycles everything	-wasteful, linear growth
-illness is rare	-health is rare
-self renewing	-doomed to collapse
-equality for all	--imprisons and enslaves
-true to itself	-delusion of separation

The genius of wild nature springs forth from a chaotic order that defies rational thought and can never be indoctrinated. It thrives without being controlled, making lawmakers and engineers not only unnecessary, but destructive to our world. A being that lives in touch with it's source, exists in a state where it's actions bring joy and health to itself as well as to it's surroundings. The physiological processes of earth create clean air, soil and water through the actions of its inner beings doing what they want. When our free will coincides with the will of our world we live in a state of grace.

Empires seek not only to domesticate the land and its inhabitants, they also impose their will onto the temporal dimension. Conquistadors of time fracture the eternal flow into minute pieces. Clocks incrementalize the day, creating the divisive delusion of the uniformity of time. No mechanism can accurately judge the passage of time, as they slow down or speed up according to the perceptions and emotions of their observer. The clocks' presence alone is a ticking coercion. Implied in its mechanism is that time has its own absolute and orderly existence outside of the experience of Mind. Every being lives at its own speed. Time is relative to consciousness and metabolism and can never be standardized. A lifetime passes through the eyes of a mayfly, while a sloth paces its way up a tree.

The cordoning off of time is equivalent to the partitioning of land, erecting barbed-wire borders where their once was fluid continuity. Churches, corporations and armies enforce their brutal and artificial depiction of time on to free-living entities. They put people, plants and animals "on the clock', thereby stealing their time so that it may be used to create riches for those in power. The artifacts of the colonizers are filled with the life-blood of their slaves, as clothes, food and gadgets are constructed out of the stolen hours of the dispossessed. The extracted energy of earth, millions of years in the making is siphoned out to produce disposable trinkets for time-blind consumers.

As globalization levels the rich diversity of this world to make way for its monoculture Mcdonaldland, we can trace its roots to the standardization of time. The installation of global monotime on to a world with infinite conceptions of time was a prototype for the dissolution of the earths' plurality of languages, species and paradigms.

The high priests retain their power by conjuring up the illusion that they control time. They tell us when we gain or lose an hour or even remove ten days from our calendar (Pope Gregory XII). They rename months after emperors (July and August). They create temporal subdivisions, a trillion to the second. They ring the clock tower bells at specific intervals and expect us all to line up for the drudgery of work, school and prayer.

We are told that time is money, losing our innocence as we sell what once was free. Made to believe that time runs in straight lines from here to there, we lack the perspective to see that the line we walk curves back upon itself to make a circle. We fear aging and death, believing that every day leads us toward the end, forgetting that every death implies a beginning. Trapped within our myopic dystopia, we fail to see that every moment includes all others within it, thereby believing that we are bullied subjects of time rather then the very embodiment of it.

Creation cannot fully remain in heaven, as all evolution would cease. Change is dependant on disruption and imbalance. It appears that we have been placed into a time in which the forces of disparity seem more powerful then the dream of heaven. The will of death has overshadowed life. The idea appears that maybe we are meant to demolish our world, so that the next era of being can commence. If so then it is a mighty strange situation we've been born into. We are given a choice as to how we are to lead our lives. If we raise up the Destroyer within us, possibly fulfilling the earth's destiny, we would ultimately annihilate ourselves.

Someday when health is restored, disease will only have a minor part in our world's drama. The ideas that materialized into the vast infrastructure of conquest and empire will fade away, as the ruins of what we once were will become a fossil layer in the planet's crust. In a few million years an observer looking into a ledge of exposed rock will suppose that a terrible comet plunged to earth, forever diverting the course of evolutionary flow.

The parts of ourselves that can still love, imagines the next era into being, even as the sleepwalker within us stumbles towards annihilation.

It seems that the greatest gift that civilization has given to eternity is its drama, the greatest contribution is the story itself. It is a contradictory fable, a hero's journey, yet also a steep fall from grace. A tale of a hyper-evolving being audacious and arrogant trying to appropriate a world that is inseparable from itself. It is God blinding itself in regards to it's own identity. Billions of stories told each reflecting a unique light shining from it's own angle. Much of our struggles are full of absurdity and obsessive madness, creating both humor and tragedy. Our cities are massive stages displaying endless spectacles of cops and robbers, husbands and wives, vehicles and pedestrians. Within the city a huge diversity of curious creatures carrying fantastical gadgets, wearing multiple faces and interacting in a variety of situations. Each city is the overmind of its varied inhabitants, a vertical luminescent hallucinatory stage set. Even the horrors of war and genocide, the extreme violence of civilizations veiled shadow give life a powerful intensity born of fear. We're all tricksters, perpetual fools working like crazy trying to build a utopia that already exists within us. Congratulating each other for discoveries we already know, but have long forgotten. We've spent thousands of years arguing with each other, trying to decide how to live when the answers exist outside of words and evaporate in the chatter. We await the messiah that has never left us. We tirelessly battle with nature that at core is our soul. Our gift is our story, for all of our actions and thoughts will live forever in time.

Our technologies are both, an extension of our bodies and the connective tissues that fuse us into a larger organism. We attach flippers like an otter, wings like a swallow, eyes like hawk, and nose of a hound. We summon the spirits of all the worlds' beings, and through our creations we make our home in all habitats. An appropriate life for a being that's not sure who they are, we rehearse all forms and actions to see how they fit. Our machines are shiny insect shells waiting to be animated with our presence, increasing our strength a thousand fold. We tunnel through the ground, leveling mountains, filling lowlands, shredding trees and building our structures in the created space. Not adapting ourselves to our world, but using our new- found strength to bully the land into submission. Our food processors are extensions of our teeth and stomachs, just as our sewers are intestines. Machines and factories eat, drink, breathe and excrete; mutated macrocosms of our fragile bodies. Our cameras are eyes, always watching, assuring us that the world still exists when we go to sleep. Our clocks are anchors, keeping us solid and mired deeply in the illusion of time. Merged with our inventions, we are cells and organs of the greater Human body we've brought to life. Seen from above, our interconnected meta-metropolis hints at the alien over-mind that has put us to work. Cars speeding down highways are blood cells pulsing down arteries, exiting onto capillaries and delivering nutrients to waiting cells. Trucks and trains transport materials between organs. Power plants burn the earth's stored energy and pump it into a million miles of rubber insulated, copper axons. Office drones in cubicles, thumping keyboards are neurons in a global brain, sending pulses of energy, stimulating responses in other lands.

Our inventions are an externalization of lost inner powers. Telephone replacing telepathy, diagnostics replacing intuition, vehicles taking the place of astral travel and television filling the void where our dreams used to be. The enormous gap left behind by our forgetting of universal wisdom and knowledge is partially bridged by libraries and archives. Our spiritual unity is replaced by a material coherence as computer networks burrow wires into our brains and bodies. Global positioning satellites and compasses replace our inner sense of directions. Calendars are needed as city lights obscure the stars. We've created laws to replace our innate moral sense, as doctrines and ideologies subvert our conscience.

With each problem that technology seems to solve a host of new ones are formed. It is a Growth industry. As people grow ill and the earth degraded, technologists offer up an army of demons disguised as angels. The dream being sold to us speaks of a future where diseases are eliminated and aging and death are finally cured. Mankind will live in peaceful harmony thanks to vast and interpenetrating communication networks linking all of fractured humanity as one. Were goaded to imagine peacekeeping armies with high-tech gadgetry on humanitarian missions. Insanity, depression, boredom, anxiety and shyness controlled with wonder drugs. People of Earth united, drinking Coke and surfing the wires. New frontiers opening on other worlds, ending overpopulation and limits to growth. Inter galactic space travel uniting "intelligent" beings from around the universe.

Behind the giddy optimism, controlling from the dark edges is the dystopian hellscape of the Machine. A world made deterministic, predictable and ultra-rational. A workforce of defeated, soulless and domesticated beasts of burden enslaved by the robot god they service and worship. A monoculture crop of sub-sapiens rooted to a phantom by fiber optic cables downloading endless streams of pre-digested data info-tainment designed to produce apathy and passivity. Electro-digital cables creep across mindscapes lassoing alienated cyber serfs foolishly seeking a unity of separation. A false messiah of Globalism transplanting mother Earth as our larger entity.

A century of high tech butchery has resurrected the Ancient Roman ghost of the Universal State in a new body. A fantasy global village brotherhood interconnected by Nike sneakers and high-speed downlinks, teaching the world to sing corporate jingles, waving flags of sacred trademarked icons, hyping transcendence via sweet black liquids.

In the shadow of this scripted mythos are acres of office cubicles filled with plastic -tapping convicts, frantically taking dictation from pseudo-illuminati, tri-lateral, neo-liberal, proto-fascist wardens. Here spontaneous culture is replaced with a mass-produced disneyfied caricature. Envision B-movie actor\dictators, caricature Muslim villains, president seducing whores, murderous sports heroes, collapsing towers, and kiddy gun slinging massacres. Is the Peter Pan pop star a pedophile or a persecuted pied piper or is he, more interestingly a monarch mind control mole who is playing out his programmed part? Picture a story of the unthinkable, about a president of an empire who helps plot a colossal attack against his own country completing a fascist coup seventy years old in the making.

Imagine pain-free vicarious living through sky-beamed demi-gods, authenticity for sale to the highest bidder, personality reduced to a convincing façade, character a quaint novelty and nature a theme- park amusement.

The great Usurper learned to project his stories, a philanthropic gesture to the duped paeans of the Mind. Many years in the waiting he demanded his price.

"The tales I tell will envelop your dreams,"

With idols cemented to the highest places, he enclosed all frontiers and incorporated a monopoly on thoughts. The minds' ocean was cloaked in a drift net. Conceive of a billion children pushing at the frayed boundaries until exhausted and carried on its wake. The shores of light, which once were a refuge for tormented mystics found itself gentrified into a mine field of razor wire neurosis.

One day the war appeared to be over and nations opened their borders. The Usurper had changed his tactics and freedom was redefined. As the dreaded curfew was lifted hardy souls were given access to the darkness. Quantum tendrils of curiosity and lust entered into the hidden places of night. Tombs, vaults and dungeons turned self conscious by the hungry retinas of the watchers. Once shameful addictions became market commodities in the black market of hoarded electrons. Those that entered hardly paused to reflect on the Usurpers new spirit of openness and generosity.

At the zoo, soul wounded creatures slipped the gates of their climate controlled existence, only to practice government sanctioned sin before corporate spy cams. What from a distance appeared to be captives freed, resolved itself as legions of fattened chattel oblivious to the transponder tags stapled to their hides. A massive stage set was erected the size of the Earth. It served its purpose as the production studios for Realityvision, but who in their right mind would have guessed that it's reason for being can be traced to a laboratory experiment in human analytical psycho-morphology. The production studio was assigned to stage world events and to broadcast real-time theatrics, meanwhile the laboratory analyzed all transactions in response to fashioned traumas. Even Sir Isaac, blood tinged in mercury, cringed at the stated goal of total date omniscience.

The financers of Meglomaniacal Ventures helped pen the storylines. Most wars were underwritten by these barren robbers just for their shear entertainment value. The opposition decried the resurgence of an American empire not realizing it was a just a simulation performed for the outrage of the Europeans. The Chinese were handed the baton and they frantically ran on fumes towards the finish line. The Chosen Ones were baited and blamed once again, asked to play Jesus for one final performance.

Minds were sectioned in order to find the root molecule of desire. A minion of masturbating males were monitored by cubicled timekeepers. The idea was to gauge a subjects perversity to the millisecond. Images that gained maximum exposure in the visual cortex and which in particular climaxed the session were most analyzed. Knowledge gained was quickly employed by the applied science of advertising. Truth seekers and rebels were handed stacks of classified documents and each trespass was duly noted. Subversion was legalized for those that revealed their deviant mental pathways to their captors. Mechanized trendspotting parrots were employed to standardize novelties as pulp writers milked diversionary scapegoats. The stage was set for condition red as the Usurper took the directors chair.

Spewing from boardrooms and think tanks, mass produced culture is endlessly pumped into absorbent minds. Those in power choose what we argue about, scripting both sides in meaningless diversions. They decide what styles are hip, who we are supposed to hate and who that we adore. Refusing to shut down our televisions that feed us our identities, we are manipulated in any direction. We believe in absurdities such as humanitarian bombings and underdeveloped countries. We've accepted that computers are our future, and that the "environment" exists outside our house. Both sides of every argument that we are allowed to debate are corrupting. We have pro-life vs. pro choice, Darwin vs. the bible and democrat vs. republican. We are allowed to argue that the government doesn't have the right to tell us what to do with our bodies when it comes to abortion but not when it comes to drugs. We still insist that the United States is a free democratic country rather then a totalitarian corporate puppet state and a brand name symbolizing, but not delivering liberty. Every new idea becomes tainted and merchandised. Great minds are auctioned to the highest bidder. We've learned to ignore that the heads of production, the disseminators of information and the political leaders are all the same people. Even alternative media is saturated with planted disinformation, channeling honest dissent towards scripted conspiracies. We pretend that our system is a democracy rather then an efficient way to get the most greedy and ruthless people into power. We don't like to admit that the chemical companies that make us sick are merged with the pharmaceutical companies that treat us, profiting on both ends. We've accepted that every aspect of our lives are recorded, monitored and graded as part of our permanent record.

Somehow we have repressed our ability to care for the future of our world and have relinquished our power to those that profit from its exploitation. We're encouraged to fight any part of the system, but never the whole. Exhausting our energies as the system quickly learns to bypass our reforms. We're herded into categories, then internalize our labels and defend our positions that we are handed. Left-right, black-white, catholic-protestant, our paths are guided, our words scripted and already spoken, the trap is sprung. Revolutionaries struggling to be free soon are marketed and incorporated, another face on a many-headed beast. Egos blazing with new power, leaders corrupted with violence, the hands of the masses bloodied. The idealist son becomes the cynical father by making war on his empire. An eye for an eye blinding everyone.

Emerged into the gaps of the worlds' imagination, I asked what could be known. Not seeing that sight transforms the seen I naively believed that my theories were being confirmed. How could I have known then that the evidence was being sculpted by the strength of my own convictions.

Crashing upwards through the domed city, I stabbed into space. I begged for the gravity of objectivity and the retention of the logical consistency of the world, a thought paradigm that would return my sanity. Slowly the soul experienced its contractions and the sun dimmed to another star. I wondered if a universe composed mostly of dark matter and empty space implied its malevolence. Answers arrived from below, as legions from the demonic order entered the theater right onto the stage. Chased through the smoldering ruins of Fallujah, or was it Grozny, I then knew that the Nazis had won the war. The leaders had eyes that gave them away, alien orbs in an otherwise human face. Around corners appeared supermodel succubi and giant insect replicants. For a moment forgetting the growing zombie mob my reflection returned to me from the last unbroken window.

I partially awoke, my cracked head within a catscan bonnet, arms and legs in restraints. Each blast of invisible rays drummed the ghastly image deeper into my memory. My eyes were like theirs!

The pen name is a brutal act of deception, both a symbolic dissection of my being and a patsy ego that can feign knowledge. In what other ways might I have deceived you? In lying to you and pretending to be totally sane, perhaps I stroke the ego of the antichrist by playing a savior who can't save himself. Is it an obscene act to mimic a messiah, but from the shallows to be dammed?

It is only through my mask that I can claim to be speaking truth, for who wants to hear of my true ignorance. Maybe, from the voice of my alias I could create a fiction, a face to the world of an unlikely hero persecuted by mechanized authority and martyred for seeking truth. Who would dare point out my hypocrisy when I could blame the character and say it is he, not I that would deceive you.

In a foolish experiment, man has been dumping tons of carbon dioxide into the atmosphere. The process of burning fossilized life forms that have been buried by the earth has been greatly expanded as many new countries have joined the suicidal mission. We have seen the hottest years on record and the destabilization of weather as a result. As forests are burning, ice caps are melting, and sea levels rising leaders continually prod us down this disastrous pathway. The weakening of the earth and sickening of its inhabitants are not just a byproduct of war but a method of conquest.

The human body contains within itself the clues to its own ancestry. Our past stares back at us when we look into a mirror. Our grasping hands, forward facing binocular vision, and rotating arms testify to our former life in the trees. Our non-primate features such as upright stance, small canines, massive craniums give clues to a more recent and mysterious past. As the fossil evidence is vague, interpretations are varied. All perhaps have an element of truth. Some believe that we are descended from aliens. Our unearthly creations and our alienation from our world seem to support this hypothesis. Perhaps we resonate with beings in another world and our evolutions converge. Aliens need not travel in ships, as mind provides easier flight. There is a theory that claims that people have evolved from aquatic apes. Are our upright stature, hairless skin, and down-turned nose vestiges of a watery past? A smooth coat, subcutaneous fat, and a shifted pelvis are changes that have occurred to dolphins, seals, and penguins when they morphed into sea creatures. Maybe we did the same. This theory is bolstered by the rarity of swimming primates, an act we do with ease. No fossil evidence exists pointing to an aquatic ancestor and thus it remains a tantalizing possibility. We search for the missing link between apes and man neglecting to consider that chimpanzees may have descended from ancient humans.

Many people have spent a lot of time arguing what our nature is and whether or not we have one. To unite these explanations is the idea that we have all natures. We have the ability to host any archetypes we choose and in any combination. We can create or summon spirits in which to conform ourselves into. Our bodies respond to our multifunction, multipersonality mind by taking on a hodgepodge of features evolved to suit our open nature. Some of our features such as bipedelism, vocal range, monogamy and flight are avian in nature. Our civilized social structure, which is foreign to other vertebrates is strikingly similar to the world of social insects, such as ants and termites. The structures of tribal cultures are often closer to those of other primates and includes the loose hierarchy of chimpanzees, the matriarchy of bonobos, the hermitage of orangs, the harems of gorillas and the monogamy of gibbons.

Most of our physical features as well as our brains lack any particular specialization. Our teeth can eat many types of foods, our hands adjust to any task, our senses are highly adaptable and our musculature and skeletal system can change to suit a wide variety of lifestyles. Our multi-purpose body evolved to suit our any-natured mind. Our brains complement our minds by rearranging themselves according to our thoughts. This versatility is not absolute as the tendency towards any specialty is inherited. Each half of the brain is involved with traveling in a different realm of reality. The left side deals with the more shallow level of material appearances. Its functions are for analytical and empirical tasks dealing with the physical world as well as exploring the more abstract mathematics and logic that structures the collective mindscape. The right side of the brain deals with the deeper, spiritual reality behind the order. It allows us to transcend the rational and explore the creative, intuitive and timeless mystery.

Our any natured brain and body reveal that we exist in an evolutionary limbo, a creature that is searching for a home. Perhaps being an eternal searcher is an archetype in itself, a being that mimics the universal substance by morphing into all things. Maybe we are just incomplete and still in an undifferentiated embryonic state. Without an obvious center to gravitate to, our nature almost appears to be hit and miss.

Our evolution is being engineered. The universal flow within nature that harmonizes evolution has been diverted and dammed within us. The checks and balances of natural order have been removed from our clear-cut, climate controlled existence. We have become domesticated.

The evolution of dogs, sheep, rice and many other beings have been so heavily influenced by the mind of Man that they have trouble living in their free state. We consider them domesticated as opposed to wild, as they have been altered to the point where they are dependent on us for their survival. Humans share many traits with other domesticated mammals such as passivity, blind obedience to authority, servitude, lack of hair and high variability of colorations and forms. Most of the traits, by now have become inheritable and are passed on to our young. Some wildness still remains. The spirits of children often have to be broken like horses before they are become willing slaves. Many people still struggle against submission though there are few places left to run to. Many domesticated animals have neotonic characteristics. They retain juvenile traits into adulthood. Through artificial selection of animals with slower development, juvenile features can be produced in adults. The domestic pig lacks the hair, tusks and much of the free-thinking abilities of wild adult pigs. Humans too, have been reworked into big, pink babies, made helpless and unable to function away from the corporate nipple. We display the traits of infant apes, with similar body proportions, little hair and an enormous head relative to the body. Artificial selection implies a choice made not by nature, but of something else. What is it that is possessing us? It is something that rules from the shadows, as we are unable to see it. It clearly isn't something that has our best interests in mind, as it has sucked out our will and has led us like pigs to a slaughterhouse. Earth nature seems to have an alter ego, the Domesticator, the Civilizer, a god that is a part of us as we are a part of it.

We have broken away, how and why it happened remains a mystery, home to mythology and wonder. Perhaps extra-terrestrials seeded our earth with a thought or a microbe from distant star systems. Maybe we're part of their genetic engineering project. Maybe they hide at the edges of our perception, infecting us with subtle persuasion. Perhaps our planet has been sick for some time, has lost its coherence and we are a symptom of a schizoid bifurcation in the earths personality. A half-blind runaway being traveling lost into the unknown reaches of consciousness.

Despite the harm we've inflicted the earth welcomes us back and our minds and bodies soon evolve towards a wild state. Most feral animals live at the edges of civilization but others totally break free and quickly regain the instincts and wisdom denied to them by their controllers. A tribe of people might originate from conquesting exterminators of nature and eventually weave themselves into the land they inhabit. They flow with the beauty and soul of their landscapes and sing with the same voice. When this occurs the domesticator recedes and humanity reclaims its depths, slowly losing the things that cause division such as clothes, technology, government, money and religion. These people then become an offense to the forces that separate and are targeted for termination by artificial selection. This act of culling is always rationalized as a necessary prerequisite for growth. It is the same story far and wide. The culled are labeled as the vermin or weeds that need to be removed from the breeding stock. Empires feed through genocide, distilling the vital essence from the free and healthy in order to increase power. War is human sacrifice, feeding the worst parts of ourselves, the extracted souls of the conquered. In an internal battle externalized, genocide speeds our evolution, as it extricates that part of us that prevents expansion and conquest, removing those beings that consider themselves inseparable from the earth. In the physical landscape these dissenters occupy coveted land that needs to be tamed, within our inner mindscape these "savages" reveal to us some of our truer selves, devoid of implanted layers of artifice. This part of ourselves form a conscience for humanity that mires the machine with freewill and humility. It counters competition with cooperation and integrates culture with nature.

Between the lines of history books, hides the workings of a grand conspiracy. It is a monster leviathan lurking in the untold spaces. An enormous war that tears at the tender seams of our minds. It is a war of man against man and of man against nature. The code words for terracide and genocide are thrown about everywhere; expansion, taming, conquest, imperialism, empire, colonialism, manifest destiny, jihad, protestant ethic, American dream, slavery, capitalism, Darwinism, globalism. The list can fill the index of a meglomaniacs diary. What is behind the greater trend is repressed. What motivates us remains a tantalizing mystery. Who we are becomes relative and all that remains is the vague idea that the global holocaust that we are causing is related to ego, greed and arrogance. But all of the parts fit together. The tragedy that is unfolding is by design. Someone or something is pulling the strings. A paranoid is basically a realist. His only mistake is to try and label the conspiracy and thereby limit it. To place the blame outside of oneself is to make war. To place the blame inside is to invite madness. History is conspiratorial, there are always those that secretly try to dominate and control. Those that we call our leaders are usually just figureheads. They are actors used to manipulate the masses and stooges set up to take the fall if things go wrong. Secret societies often call the shots, treating the world like a giant chessboard. They provoke one side against another, financing both sides of wars and profiting no matter what the outcome. Scapegoats are created and sacrificed to divert the anger of the dispossessed. Behind shadow governments and secret organizations is an even larger order, whose roots can barely be traced through the murky darkness in the recesses of our minds.

There exists in nature the archetype of the destroyer. An angel of death that demolishes creation and in so doing makes way for the new. Acting in balance the death force is a necessary and even benevolent aspect of evolution. Storms and volcanoes, ice and fire appear to us as merciless killers but the earth responds to their destructiveness with a creative frenzy. Death gives way to birth as the Savior is resurrected and the land is rejuvenated.

What would happen if death was repressed and feared? It would get shoved into the deepest shadows where it's power gains exponentially. We live in a culture that is scared to age and die. Death is seen as something we must fight against at all cost. Because of this, it is always with us. Death becomes warped and evil, a nemesis that we must engineer against. The angel of death when cast into the dungeon metamomorphosizes into the Devil. It is the Universal Antagonist, a being that feeds on imbalance needing it in order to survive and create it's own order. A trickster god that diverts energy from the central path and towards it's own. In the largest sense it is one with the whole, but doesn't realize it. It seeks the impossible task of divorcing itself from the source of being. It is a death force greater then death, as it seeks not just to destroy the body but also the soul. It grows branches on the tree of life then girdles them to fracture the connection. Disease and dissociation seems to be its purpose, divide and conquer its plan.

Many new-age religions often deny the evil, forcing a feel good cosmogony as a path to enlightenment. But looking only towards the light, one is blind to the shadow cast behind them. Hiding there is our greatest fears, beings of terror that divide us from our Self. The neglected flip-side of new-age beliefs is conspiracy theory, here we see forces merging together not to heal the world but to possess it. The shadowy networks of fallen beings is a travesty of divine union where greed replaces love as the connecting framework. To acknowledge one without the other serves to cloak vital threads of the narrative of which we all take part.

A terrible relativism can develop from the knowledge that all is one. Even tolerance and acceptance can be destructive traits when off balance. For how can anything be healed if we deny the disease. By integrating our runaway parts we fulfill our destiny. As far as we may have fallen, we have the power to lift ourselves up. Because of the tyranny of old patterns, and the effectiveness of mass mind control, our lives must become a quest, or else we surrender to the forces that try and control us. Our intelligence can't tell us how we should live, but it can breach our enclosure. Unlike a wild being in the wilderness that just knows how to live, we must search for the center within us to guide us home.

A process thousands of years in the making, enacted by religion, government, science and commerce has produced the ultimate lie, that our Earth is irrelevant Only with this belief can even the kindest people take part in the slaughter. Few people are willing to confront the established order, because it would mean examining their own complicity to the destruction. Evil is usually only acknowledged as the actions of others. Religions often seek to define evil as the debased action of non-believers.

Christianity's obsession with sex as sin, gives self-righteous comfort to followers and a target for hatred for those that don't comply. For some odd reason the act of destroying creation isn't considered a sin by any of the main religions. Could it be because they're followers are as guilty as everyone else?

Religions and governments seek to make laws and rules to control people. They list which actions they believe are evil, never realizing that the laws themselves may be a form of evil, as they obstruct the natural flow of things and preserve imbalance. Laws serve to hoard most of the world's energy in the hands of a few, preventing a natural balance from occurring. As social systems become more indoctrinated and beurocratic pathways are chosen for us, and freedom is sacrificed to the gears of the machine.

Among mystics and heretics a more open idea of evil is acknowledged

Meister Eckhart –"falling off from the good of nature", "contraction of awareness"

Thomas Aquinas- "An action that obstructs grace"

Norman Pittenger- "estrangement from self, others, God and nature"
"misuse of human freedom to destroy life, spirit and love"

Abraham Hershel- "the refusal of humans to become who we are"

In Taoism evil is lack of balance and the refusal to follow the spontaneity of nature. No action can be said to be good or evil except in the context of which it is done. A wild land and a healthy body exemplify the innate wisdom of our universe that brings creation towards harmony. The existence of man-made laws deny our ability to know right from wrong and therefore based on a negative view of humanity. We are told that laws prevent us from killing each other, implicit in this is that we are killers. Then we are encouraged to slay each other in government sanctioned warfare and crusade.

Natural law cannot be written, any dogmatic interpretation is bound to fail, as justice is contingent on time, place and circumstance. Order is not maintained by rational means as existence is governed by synchronicity, pervasive mystery and laws of paradox. We make a mistake in assuming that even the physical laws such as gravitation and speed of light are stable and not subject to evolution.

It seems that whenever the System gives you a choice between two beliefs, they're both screwed up in some intrinsic way. Communist-Capitalist, Democrat-Republican, Darwinist-Bible thumper, pick your gods, bet your favorite team, sculpt your reality, dig a trench, go to war. A carefully scripted revolution, whether you win or lose, no matter how many you convert or subvert your still messed up. The dice of revolution are loaded. It's not a battle of opposites, but a competition of similars. Put on your masks and stare into a crooked mirror, strange how left and right point the same direction. Monarchies, dictatorships, democratic republics, communist regimes, fundamentalist theocracies, fascist corporate states, chiefdoms, scientific technocracies, socialist welfare bureaucracies, alien puppet states; in all their differences they all seem to evolve into a coercive rule by an elite few over legions of slaves. No matter how much the zoo tries to recreate naturalistic habitats, the monkey still sits in a fuckin' cage.

This is a work of propaganda. Baiting with shards of truth and hooking you with lies of omission. At the extreme edges of thought and ideology are entangling snares of paradox. Belief is ever divisive, as it implies the absolute existence of the believer. Even as a voice in your minds' eye speaks these words, this book sits like an object in your hands, the paper believed to be not of your own flesh.

Billions of dollars are made by the pharmaceutical/ psychiatric racket to return people to the reality system that has been engineered for them. Psychiatrists act as priests, saving heretics and burning infidels who refuse to recant. Few are true healers, genuinely attempting to make their patients more whole. Taught that most mental disorders are chemical imbalances, doctors treat the symptoms, seldom realizing that the depression, anxiety or mania are signs of a mind attempting to heal itself. Living in a dissociated state is the accepted norm in our society. People spend their lives building walls to keep out the depths of themselves, bullying their minds into remission and using their energies to support the empty personas they've created.

We worship the ability to be false. In time, we are taught to play the game, to act professional by separating our personas from who we really are. Those that are unable to sustain the deception are characterized as drop-outs and losers. To accept the label is to fill the lowest rung of society. Rejecting the label, one can see that their abnormality is a gift. Without the barricades of society fully in place wide horizons open before us.

For those who have a great ability to feel and whose conscience is strong, living in a divided state is difficult to maintain. Holding on to the imbalance results in neurosis, as the denizens in chains struggle to break free. Their illness is a positive sign, proof that their soul is still living, that they haven't yet been totally broken. The pain that is felt is a natural reaction to the state of the world. An ability to feel is not an illness that needs to be drugged into submission. Those cast out by society as inferior can take pride in their inability to function in an insane culture. We worship the ability to be false. We are taught to play the Game, to act professional to separate our persona from whom we really are. Those that are unable to are characterized as dropouts and losers. To accept the label is to fill the lowest rung of society. Rejecting the label, one can begin to see that their abnormality can be a gift. Without the barricades of civilization fully in place our horizons open up before us.

For those with the deepest ability to see, the prescribed reality becomes an absurd façade, living in a world of lies drives many to the point of breakdown. Some survive by expressing their truth through art many others fill the asylums. In more enlightened cultures they might have been shamans, mystics or healers. Here they are labeled schizophrenics, then drugged into acquiescence. A deep fear of higher states of mind is instilled in people as it is equated with insanity. Our cultural paradigm doesn't allow for the experiences one has in these states, and can't even begin to explain what is felt.

The seer of truth is a great danger to a society built of lies, and if made to fear, he is preyed upon by demons who seek to break him down. A process of psychic annihilation is put into motion by labeling a seer as insane. Patronized, then ostracized for his strange behavior, doubts and fears attack from within. Without teachers to guide him, he is left at the mercy of doctors, who cannot comprehend his experience. Without a world-view that includes the truth of his experience, he is lost alone at the edges of a forgotten realm without a map.

The outsider, besides feeling alienation from society and the pain that goes with it, gains the universe. Infinite pathways that were denied from within society, open up for those that travel outside the boundaries. The system channels many toward the archetype of the criminal, the violent lunatic, the drunken fool, or the homeless beggar. These types have been scripted from within, serving an important role in the establishment. They provide confirmation that uncivilized man living without laws degrades into a terrible state of being. Over and over we hear the cliché "like an animal" to describe murderers and rapists, never giving a thought to the state of grace that wild animals often live in. Words like anti-social or sociopath are used to describe the selfish, disconnected narcissist, ego-driven killer, not realizing that this describes not the true outsider, but the unmasked civilized man. In the powerful shadow of the civic minded, business man, father figure is the serial killer, hater of women, killer of natives, destroyer of earth. The serial killer isn't man stripped of civilization, but civilization stripped of pretense. Not the Animal but the Alien. Not man's true nature but nature repressed and mutated. What else can explain the evil of the ordinary, the hidden side of the family where fathers beat and molest their own children. The spurts of genocidal fury carried out dutifully by law-abiding citizens, who after the killing spree subsides resume their regular life. Our civilization, our religions and our families are not evil, but they have served to nourish the killer, shoving him deep down inside and out of sight. Always projected outwards, but rarely seen from within. For if he was known, he would be diminished. He is the architect and engineer of the machine, whose power lies in his elusiveness and ability to mimic God.

The criminal is not the outsider, outlaw of society that we are made to believe. He is a freelance business man. For what is capitalism if not armed robbery. Why is a billionaire robber baron who hoards the worlds energy from the starving, looked upon as a hero? Meanwhile a homeless man, stealing food is locked away. Stealing is legitimized if done to the poor, rape is legal when done to the earth. Capitalist democracies, communist regimes, totalitarian fascist states, corporate empires and dictatorships differ mainly in their methods of control. All eventually become the same. A small powerful elite controlling the minds and enslaving the bodies of the masses. Within each system people are told that the control is necessary. Every leader insists that the people are free.

In a wilderness there is no theft, as it's not possible to take more than needed. Without armored cars, Swiss bank accounts and police protection a balance is reached. Possessions are temporary and everything is free for the taking. Hoarding of resources is pointless, when detached from abstract qualities such as money or power.

Theft is a result of a culture trying to balance itself. When a few people have nearly everything and the rest are left with crumbs, theft is inevitable. The situation is complicated because peoples' desires have been hijacked. We no longer know what we need only what we are told to want. Our cravings and wishes have been engineered to be endless.

Outsiders play an important role in our culture. They are the fountains of creativity. From within the walls nothing new can arise. Empires must constantly evolve or else fall to other kingdoms. Arms races need creative minds to imagine the technologies of the future and new methods need to evolve to control the people struggling to be free. Minds trapped within the fishbowl of society, have trouble connecting to the universal flow of creativity. Every system needs to evolve a way to reel in the dissidents so that there creative energy can be used by the establishment instead of against it. Money, fame and power await the creators of the New, bringing in those that have strayed. They are propped within the ranks of the establishment. Once lured in they hesitate to attack the system that has given them so much power. In turn they risk losing their soul, which is their connection to greater things. Many people are willing to take that risk. It is, of course the devils bargain. The Globalist Empire has been so successful due to its ability to make people sell out. People who once inhabited the fringes are brought in, and paraded in front of the world. The message they impart is "We are for sale, and proud of it".

Dissidents that refuse to bow down are marked as vermin to be eradicated. The public is warned of a grave danger to society, stories of dangerous sickos are circulated and labels are applied; communist, nigger, liberal, Jew, addict, lunatic, savage, radical, degenerate, terrorist. Waves of fear are fanned across the land until a siege mentality is born. Attacks are provoked or even staged and an unbearable tension is produced. New laws and government agencies are then created to deal with the menace, civil liberties are denied and the noose is tightened. The story is always the same.

In the late 1960's many people were able to temporarily break free, as the system was cracking under it's own beurocratic weight and lack of conscience. The brief period of openness and freedom didn't die because of the mass arrests and assassination of radicals which were commonplace, but because the radicals were reabsorbed by the evolving system. Society itself temporarily became more open and much good was accomplished. People genuinely attempted to heal the damages caused by racism, sexism, and warfare. We began to become conscious of the massive destruction we were causing to our planet. The system quickly adapted to absorb its deserters. Soon enough freedom, peace, and love were stripped of meaning and reduced to the emptiness of a corporate designed fad. The gears of the machine were given a day-glo, tie-dyed, paisley, have a nice day paint job and a new and improved user friendly control panel and hippies hopped aboard. The ability to manipulate and influence any cultural movement is endless as long as the public sees its reality from a T.V. screen. Mass mind control through media technologies has surpassed even the imaginations of science fiction writers.

The Corporation has infected our inner world. In the apparent absence of earthly or heavenly revelation, the corporation has become our surrogate parent, a macrocosm for our mindscapes. The hierarchal structure is a legacy handed down by the Nation State and before that the Church. It is well entrenched in our being. The corporation is the latest form of government and pretends that the earlier forms still have power, manipulating feelings of nationalism or religion to further its goals and wars. It is the materialization of beliefs in a machine patterned universe and cutthroat Darwinian competition. Efficient, amoral, greedy and monopolistic, the corporation expects the same of its component workers and consumers. It has melded technology and flesh to birth its progeny cyborgs.

The corporations' physical aspects and assets are hard and objective empowering ideas of solidity of matter. Office towers of glass, stone and steel reveal their greater design from a distance; colossal microchips plugged into city–scale circuit boards. The wiring crackles with energy hacked and mined from the earths burial grounds. Electrons pumped through circuits, interface with our senses and synchronize their flow with the electrical discharges of our central nervous system. We bind our emotions to numerical peaks and valleys of massive pyramid scams desperately trying to avoid depression. As the analog eccentricities of our minds are phased out, digital determinancies are implanted. Our brains have become research and development laboratories for artificial intelligence.

Hovering above the earths' surface the Corporation is rootless, seeking out pockets of life-force to siphon out. It's human resource travel where opportunities abound, meanwhile its material center of operations constantly relocates to places where people sell themselves cheap and energy exploitation is easy.

The Corporation employs a split ego format on its structure and imposes it on its modular humans. Its face to the world is sculpted by teams of psycho-engineers working public relations. The corporate persona reveals a being that is benevolent, efficient and polite. It is smiling, sane and trustworthy, speaking in words measured and controlled. It is the professional mask of customer service personnel, politicians and executives. It is the face we show others in order to get a head in the world. The image projected by the Corporation casts a powerful shadow, there, in the darkness lies its true motivations and methods.

The corporate mask is an amulet that protects its user from prosecution. Endless war, animal torture, wage slavery and planetary plundering are acceptable vocations when done with the correct attitude. This differentiates the Corporation from other forms of organized crime. Imagine the outrage upon hearing that a terrorist organization was to poison the earths waters, yet General Electric thrives despite its sickening the Hudson and Exxon functions with impunity following its fouling of Alaska's' pristine shores. Weapons of mass destruction are evil in the hands of rogue states, but just another consumer product to the manufacturer.

In the hidden passages of our inner world we hide our complicity. On the surface, seldom is discussed our part in the War. As if in secret alliance with our corporate handlers, we can't see the sweatshop in the sweater and are blind to the house of horrors in a canned ham. Dutifully paying our taxes, we don't acknowledge the nuclear missile that we purchased. Our blindness is the true price we pay for the façade of normalcy worn in a time of insanity. We can't see out because we refuse to see in, too scared of what we've become.

With all I's diverted to Jerusalem, I set up shop in Shanghai. I am the Executive, now commanding an army of 5 million teenage girls who assemble the worlds products. I thank The Abusive of All Fathers who rendered each one of them docile and subservient, but now lies catatonic in the Old Folks Home. I pierce the heavens with spires so that milk and blood can be siphoned off and reprocessed into endless spectacles of nauseous melodrama.

I am the Executor, reading the last will and testament of what most would consider an American birthright. Through you children of China, I free myself from the burden of empire. To the great globalists you were a generation of Mao's' amnesiac angels, pupas of capitalism forcefully turned to gaze away from the ring of fire breaking the night as it spread along your eastern shore. Now you run towards Mammon somehow carrying a messianic optimism while breathing toxic air and answering to mobsters. I fear for you as you have cashed in your forests for veneered particleboard furnishings that we browse at Wal-Mart.

I am the Exorcist, I attempt to banish my demon by passing it on to you. I delight and regret to inform you that your Gang of Four are now climbing the upper rungs of the corporate ladder and are excelling at the art of Subtle Invasive Persuasion, having long since perfected Pointless Beurocracy and Vindictive Thuggerry.

I am the Executioner, beheading your captive slaves and selling their organs on the commodities market. I've drowned your ancestors in the chemical sea. The Dragon has been broken and is now impounded at the three gorges.

I am the Expatriot, my allegiance is to no One, my migratory routes closely follow flows of capital. I thrive where the land is pliable and the people ripe for harvesting. Once your will and your habitats have been acceptably devastated, I abandon your atrophied world and take to the air seeking new frontiers to plunder.

I watch the emergence of legions of newly urbanized Kulaks. Eyes newly rounded and femurs stretched, you are the ambitious executive replacements. So what if you commence your reign in a world already spent. Nobody told you that your skyscrapers are anachronisms and that your new cars are powered by a vanished fuel. So what if you have joined the festivities a little late and the revelers are wasted, rejoice in the hostile merger that has ended the cold war and go on such a binge that your promised salvation will arrive a hell of a lot faster.

The panopticon is being assembled. Quickened by the ascendancy of an ever more fascist regime, the mechanism for total surveillance is being put into place. Jeremy Bentham wrote of the idea of the panopticon, a prison where each inmate is potentially under the watchful eye of the warden. In time the coercion is internalized and the prisoners are incarcerated in their own bodies.

Welcome to the Global Detention Center, where all of our actions are potentially observed, our transactions monitored and graded, and our minds reformatted into the narrow bandwidths of cyber circuitry.

Totalitarian data profiles overshadow our real personalities, tracking the foods we eat, books we read, movies we see and places we travel to. Invasive technologies now exist that would make Stalin salivate. The system watches and records, as we unknowingly leave behind a digital slime trail, that ripples with our every action. Cameras on street corners never sleep, their tireless eye encoding our image on magnetic tape, or else downloaded to facial recognition programs, where our presence is noted. We try and tune out the spying cameras, but their presence is still felt. Dreams of freedom dissolve away as a platoon of digital spies monitor our every movement.

Children are prepared for a future dystopian cyber-prison. Their school is their proto-penitentiary where they are scanned by detectors, are fingerprinted, mentally strip-searched and have their urine analyzed for forbidden actions. They are subject to constant internet scrutiny, as corporations buy and sell their profiles to their exploiters and jailers. Children enter into a world of statistical surveillance, stroboscopic photocopy, risk analysis, motion sensors, spyware, infra-red wiretaps, satellite tracking, holographic projection, global data banks, tempest technology, listening posts, voice recognition and electronic flagging. This is the world of micro-engineered paranoia.

The most powerful tools of the new technocracy are the ones that you never hear of. What is the outcome of 40 years of CIA research into mass mind control and the programming of assassins? How are the results of experiments into controlling the minds of the masses with radio waves being implemented? Do we realize how easily a plane is knocked out of the sky with an electronic pulse? What is the fruition of research into remote viewing, anti-gravity, time travel, teleportation, death rays, climate engineering and scalar weapons? Do we really want to know? Hiding in the piles of disinformation are the means of control in a future of inevitable chaos. We've allowed governments to rule in secrecy. How couldn't we see that they'd use their power to exploit and oppress us?

The most fearsome of our created nightmares is the specter of biotechnology. As those in power unleash their mutated monsters all over the world, we have become helpless spectators and clueless lab animals. The marriage of science and capitalism, seems to be the product of a demonic matchmaker. Their offspring are a million mad scientists releasing their destructive inventions in an endless search for profits and conquest. The world is their petri dish and we are their subjects.

The completion of the human genome project is a proud milestone in the quest for knowledge. But knowledge without wisdom is deadly. The genetic engineering of humanity will bring with it unspeakable horrors. The last line of protection for humanity will be to deny access to our blood and genes. Race specific bio-weapons will become outdated, when virus' can be engineered to kill specific families or individuals. When those that have seized power know your gene sequence they can control your destiny and seal your fate.

Vaccinations, many of which have been made mandatory damage our immune systems and may be the main cause of SIDS and autism. They also allow the people in power to introduce various foreign agents into our blood stream. Viruses can be engineered to inject specific gene sequences into our chromosomes. The rhythms and cycles of our bodies may then be altered without our knowledge. The hijacking of our body chemistry through gene therapy is an extension of an unspoken eugenics program. The artificial selection of humans is begun in our pre-natal state. Amniocentesis and genetic testing is used to decide which babies are to be aborted and which are allowed to live. In parts of India female children are eliminated by the thousands using this process of gender cleansing. Those with Down syndrome and dwarfism are wiped out before they are born, leaving this world a poorer place due to their absence. Thousands of traits may soon disappear, silently reliving the Hitler's' process of Selektion in hopes of creating a Master Race. The reworking of our genes is a logical extension of this process.

In the near future, vaccines may be used to inject machines and transmitters into our bodies. In an age of ever shrinking microchips and nanotechnology this is not as improbable as it seems. The scientific management of reality may be extended into the inner spaces of our bodies and minds. An argument can be made that the process is already well underway.

Power is retained and centralized through the mass production of terror. Nuclear weapons have a far greater purpose then the destruction of populations, that of mass mind control. Nightmares of nuclear scenarios churning the earths' diversity into a melting pot of scorched carbon inject our daylight hours with endless anxiety, or worse a dull insensitivity. The death cloud is a hanging threat used by our leaders to instill passivity. Empathy towards our world is rendered meaningless and our emotions held hostage by weapons that are most effective when kept in their silos. Their purpose is one of deterrence, not of war but of the Real. It is deterrence through determinism, technologism taken to its logical extreme where nothing is left to chance. Life becomes a channeled stream and individuals are homogenized in the menace of a nuclear stew.

The creation of atomic weapons mirrors the formation of terrorist bands. Both involve the formation of unstable elements that are used to justify the system of oppression. Humanity when faced with horrific threats from engineered catastrophic scenarios is forced into using all available energy for defense. Vast industries grow on the backbone of armor plating our souls, with the result of being unable to feel the rapture of truly being alive. Encased in Kevlar condoms we live the illusion of being estranged to our own essence.

The use of terror to dominate people has two phases. The first being traumatization. This is to stage or provoke an event to create a panic reaction, a managed demolition that generates a call toward greater dominion. The shock to the system is a classic mind-control maneuver, the directed trauma severs established connections and the ideology of a new order is quickly implanted into the spaces of the created vacuum. The second phase is the creation of obedience through false unity, hero worship and warfare. The evil that is producing the Fear is identified to the public and a war is waged against a prescribed patsy. The created masses held in the grip of a nameless anxiety find their release in the ritualized slaughter of a scapegoat. Used in a cycle these phases exponentially accelerate the domestication of existence.

My arrival coincided with the height of power for the empire. A ridiculous tyrant was waging war on the barbarian hordes. The hordes lorded over the land of the subterranean swamps, which were all but worthless to them. However a potion was prepared from these swamps that the world-eating empire became obsessed with. It was considered to be the bottled essence of the sun and composed of the carbon of ancient dragons. The world eaters used it for everything. It was added to the food, clothing and shelters of the people. It was the power source for the artillery of mass human sacrifice. It was pumped endlessly into the air and injected straight into the brain. This potent brew was believed by the people to be the source of their libido without which the people became depressed and prone to fits of rage and nostalgia.

The emperors' illustrious family rose to power by pushing the potion on the people. They diplomatically funded both sides in the Great Slaughter II, benefiting greatly when the two powers secretly merged after the war. With the looted treasuries of the other rivals, the new team financed a vast network of colonies and puppet rulers. In time the new empire defeated its nemesis in the war of mirrors, leaving only the hordes to challenge the world-eaters.

The leader of the hordes and his family were also great merchants of the potion. Put into power by the empire to secure a steady flow. The horde leader, a closet infidel had widespread support among his people in his new battle with the world eating empire. The emperor too was popular for his eternal battle against evil-doers. What the people on both sides failed to realize was that the two leaders were actually the same man, playing the same game from opposite ends. He achieved the dual role through a complex process of magnetic projection holography. Though the disguise was thin and the acting terrible' the blinded people believed; A spell had been cast by the maker of images. By sitting on both sides of the chessboard, the deceiver was able to control the pieces. You see, the Demierge had to invent the Adversary in order to impose his false good/evil dichotomy. One day a programmed pawn managed to demolish the opposing rook. Following the screenplay of biblical prophecy, the plan hatched by the two-headed leader was well under way. Thoroughly provoked, the world eaters became enraged and attacked the hordes. The hordes in turn cut off the flow of the potion, and braced themselves for the worst. The people wondered whether they should throw the whole chessboard if the game would become too difficult to win. The ridiculous tyrant could barely contain his smirk.

Our stories are filled with characters that make their dreams come true. It is an undying fantasy, perhaps because it contains universal truth. Evolution is the unfolding of dream into reality. Reflected in our stories, from King Midas to Frankenstein to Scarface, are the unexpected and sometimes horrific consequences of when we get what we ask for. Disastrous outcomes seem inevitable for those that use their power to feed their egos and build their empires.

Implicit in these tales is the idea that we are born of a culture that has forgotten its identity. No longer being able to identify ourselves with Earth and Universe, we pretend that we are independent agents who are separate from each other and at war with our surroundings. We dream of ever larger weapons culminating in doomsday devices. We dream of endless growth and prosperity ending up with overpopulation and a ravaged planet. Our dreams, when at odds with those of our larger selves become night terrors, in which we wake from our sleep screaming "WHAT HAVE I DONE". Even minor wishes born from our wisdom vacuum backfire in ironic ways. We may wish for constant comfort and receive a device to keep our homes at a blissful 72 degrees. The result of our wish is to be always uncomfortable with the outer worlds' varying heat and cold. In a larger scale our yearning for climate control has helped imbalance the atmosphere, resulting in climactic extremes and instabilities.

Many techno-futurists dream of a coming technology where masses of microscopic self-replicating nanobots can re-engineer a pile of grass clippings into a five- course meal. All that we may ask for can be assembled for us at a single command. Everything perhaps except for meaning and reason to live.

The technologies already exist for parents and bio-tech companies to conspire to engineer children. Genes from any organism in the world can now be implanted into human embryos. Megalomaniac mommies and dictator dads can now dream of kids with I.Qs of 300 and flawless good looks. It is an easy prophesy to envision the dark outcome of this egoists' quest; genetically enhanced children feeling robbed of their freedom and souls, and their shadow selves, the horrid mutants hidden from our eyes behind cinder block walls and lime green paint.

All the ingredients for Hell on Earth can now be mixed under a microscope. Doctors working to eliminate near-sightedness might ask for a little more. Why not allow people to see ultraviolet, or have night vision or even extra eyes. Maybe scientists could produce subservient people that don't sleep and never tire, so that they can work around the clock. Why not hybridize them with chimps, so we would feel little guilt at their servitude.

The field of genetic enhancement can be an extension of product development. Fetus' can roll off an assembly line with super hearing, jack- hammer arms, glow in the dark skin, can-crushing jaws or drug sniffing noses. After being birthed from artificial wombs, each child can be raised to perform their own industrial task. When combined with robotics, genetic engineering can produce a wide array of man-animal-machine hybrids to be put to work. Furthermore they can be cloned and standardized to provide for better efficiency and easy replacement.

We allow market forces (corporations) to decide our future. In other words the dreams that we are fed are manifestations of greed. The technologies that we are made to desire are clearly not produced to create the common good, but to make a handful of people rich and powerful. A good P.R team might assuage our initial aversion to soul-robbing technologies. We may even be conned into believing that we are creating something good. Nonetheless, we can't help but to feel that we've gone down this road a thousand times before, each time with a progressively disastrous outcome.

Every molecule that we are in contact with influences and populates our inner world. Every essence resonates qualities of emotion, color, mood and personality, while they structure our world with charge, frequency and mass. Water, broccoli, gold, air, talc, mercury, plastic, DDT, shit – Their outer qualities describable in words born of our senses. Their inner aspects must be felt and then expanded by imagination, leaving only the poet to communicate their being. For most, the life within matter remains undetectable, as we are taught not to look. The universe of our animist ancestors has been placed just out of reach. Yet, in moments of awe and wonder we are gifted to really know the ocean or able to feel the power pulsing within a flame.

Deep in our bodies, in storage vesicles of gland cells, or flowing in our blood, are elixirs of transformation. Adrenaline, dopamine, serotonin, endorphin and many other potions play major roles in our minds' unfolding. They are able to shift our frame of reference and mysteriously allow different facets of our world to reveal themselves to us. Released into our bodies as a result of our thoughts, actions and ancestors, they are potions of love, fear and rapture, doorways into the hidden realm or jail cells of mundanity. They integrate our mind and bodies, and are intermediaries between our selves and the collective dreamscape we inhabit.

These chemicals also exist outside of our bodies and are produced by plants and fungi. The most psychoactive chemicals we know of as drugs. Plants create these substances for the animals that eat them and thus serve as endocrine glands of the earth. Certain plants can suppress appetite, reduce pain, give energy, heighten awareness, increase action, help relaxation, provide immunity to disease and improve the senses. Shamans use plants to leave their bodies, visit other villages and speak with the dead. The plants that enhance consciousness to the point of ecstasy serve to integrate separate egos into the planetary mind. The spirits within these plants act as portals to other plains of existence. Often they are teachers, easing the anxiety felt upon entering dimensions of being so alien to those raised on the mind-leveling consensus reality imposed on our travels. We have forgotten the magic of our world, a place where every last being or substance can provide us with insight to restore and guide us.

The greatest effort by those in power to control what we take in is the war on drugs. Those that have assumed power feverishly attempt to control our bodies and will readily lock us up if we ingest something that is not approved of. The line drawn between legal and illegal drugs falls generally between substances that empower ego and those that encourage wholeness, between blindness and increased perception, shallow and deep, obedience and freedom. The most illegal drugs are those that raise consciousness and allow the user greater perspective. Mushrooms, peyote, marijuana, acid and opiates open up the world and reveal our prescribed reality to be a narrow band in the spectrum of existence. The imprisonment of people for possession of these substances amounts to the persecution of dissidents for thought crimes. Hundreds of thousands are oppressed for the crime of sampling other forms of reality then those designated permissible by authorities.

The use of mind-revealing substances by animals is well documented, thou rarely discussed. Animals use their ingrained wisdom and intuition to find plants that help them in their journey. Some plants give them greater perception, some greater speed and perhaps some give them visions of greener pastures to travel towards. Goatsjoy, catnip, and hares lettuce are psychoactive plants named after their favorite users. Monkeys, deer and tigers enjoy magic mushrooms, to what effect only they know. This benefits the mushroom as new generations grow from shit out spores. Cannabis is spread in a similar fashion, as its seeds are carried by rabbits, birds, deer and people who partake of this wonderful plant. Similarly, morning glories and poppies attract wandering animals towards their seeds by providing them a euphoric and mind-expanding experience. The effect of eating these substances may have had a great effect on the evolution of species as new perspectives radically change pathways of growth.

Nearly every plant contains stimulants or depressants to alter the mood of those that consume them. Koalas sniff out particular types of eucalyptus leaves to raise or lower their body temperature. Animals use the vast medical chest of the plant kingdom to fight infections, rid themselves of parasites, help digestion and heal wounds. Every disease seems to have a cure among the plants only the knowledge has been lost to us.

As we travel through the mystery we have been born into, we take in the world. Gazing at flowers, eating fruit, breathing in pollen, we indulge in the essence of existence. With each intake, we resonate and strengthen that particular spirit within us, building our inner world by infusing it with the outer, each being symbolizing and acting out the spirit behind its evolution. Every plant is a personality, like us their resident chemicals are distilled spirits of various traits. When we take them in, they can become portals to different levels of being. Reacting uniquely with everyone, each drug has a general character. Like people, some are to complex to describe, others simple and to the point.

The processing of plants into a single chemical, removes many subtleties and narrows it's range of personality. The chemicals within a plant synergize to create a more whole experience. The process of making hard liquor from fermenting plants, crack from coca leaves or heroin from poppies, makes the drug more single minded and also inherently more dangerous. It removes a substance from its context of co-evolved chemicals. Synthetic chemicals can be most harmful, as our evolution hasn't prepared us for their presence in our bodies. They mimic natural forms. Their effects are sometimes novel, frightening, entertaining and unpredictable. Their spirit evolves with our experience of them. Bad acid trips were rare for many years and became common perhaps through the fear created by those in power and also due to poison mixed with the LSD. Negative propaganda changes the expectations of a drug and can also serve to dishearten the user. Criminalization of substances can induce demoralization, paranoia and fear of imprisonment.

If taken as a sacrament rather then an escape, for healing and not for just partying, certain substances can give us an entrance into the beauty and mystery that we have been denied. The creativity unleashed can open new pathways to living in harmony with the earth. Inner traveling must be done without fear and with an open mind, as our universe reveals itself to be a much larger and different place then we previously believed. Our society, when viewed from a greater perspective may appear as a ridiculous and insane place. Our leaders appear as simple minded and somewhat demonic, and our relationships with others as shallow and distant. Just like anything in life, what we take in must be in balance with our needs. Too much of anything is toxic. Drugs that ease physical or emotional pain may be addictive to those that hurt chronically. Most drugs are analogs to ones created in our bodies. Our cells may stop producing a substance if it is brought into the body for a long time. This is true of any treatments including shampoos, deodorants, moisturizers, exfoliants, chapstick, foot powder, laxatives and all other medications. Long term use, prevents our bodies from healing on their own and addiction then occurs. The health care and advertising industries grow rich by encouraging as many addictions to products as possible.

Mind-altering substances don't offer instant enlightenment, they are pathways fraught with dangers. All treasures are guarded by demons and you must prove your bravery and worthiness. Often you are confronted with your greatest fear before heights of euphoria can be reached. Integrating insights can leave you on the doorsteps of madness, meanwhile condemnation by the ignorant cause pangs of alienation and sweeps of incarceration. The fury directed at those who grow this way reveals the fear that is felt by those who insist that the world should only operate in a particular mode.

The massive amounts of anti-depressants pushed on us by experts, reveals the use of drugs as a form of social control. Increasingly, the elderly, children and even pets are being drugged for their inability to cope with a world gone mad. Emotional pain and depression are treated as merely chemical imbalances, rather then symptoms of living in an unnatural state in a pathological society. Compassionate and sensitive people are labeled as chemically defective, then drugged so that their powerful ability to feel is reduced. With their empathy sedated they can then become productive units of society and not have their conscience get in the way.

Prozac is the elixir of conformity, blocking emotional pain and erecting barriers to insight. Certain personality types are being targeted for extinction. Shy, melancholic, introverted, contemplative and intuitive people have trouble functioning in a cruel and cold environment. The insight and rebellion that they spawn are a threat to power structures. Prozac is a killer of activists, limiting societies ability to heal itself by sedating the seers.

Demons	**Pharmaceuticals**
Valefor	Prozac
Zepar	Strattera
Focular	Xigris

Oriax
Vapula
Zagan
Volac
Phenex

Zyprexa
Zolaft
Zyrtec
Viagra
Celebrex

by Tim Boucher

Food is packaged thought. Symbolically and literally we are what we eat, ingesting, digesting, metabolizing, incorporating and eliminating our surroundings. Our society's ability to influence and control what we ingest is a form of mind control, not at the level of information but of comprehension. We are loaded down with foods that numb us out, dumb us down, blind, deafen, desensitize, narrow and solidify. Book burnings and censorship are largely unnecessary in a society that has become deadened to truth.

A **look at ways that food is desouled**

* separating people from the land they live on (urbanization)
-destruction of naturally occurring food source (deforestation)
-discouraging people from eating wild foods (mushrooms, nuts, herbs, fruits)
* pollution of land, sea, and air
- fluoridation and chlorination of water
- making water undrinkable at it's source
- toxifieing all life forms with industrial waste
- inducing climate instability
* encouraging importation of food instead of locally grown
-setting up economies for export
* Agribusiness (mass production of food)
-toxifying food with synthetic chemicals (pesticides, dyes, preservatives)
-toxifying animals raised for food (hormones, pesticides, antibiotics, horrid living)
-deadening food with poor soils and chemical fertilizers
-genetically altering plants and animals

-eliminating diversity through monoculture plantings
* processing of food until it's biologically inert (white sugar and flour)
-mechanization of food production (elimination of art and soul)
* making food bland and predigested
* pushing meat, dairy and eggs by government agencies, advertising and P.R.

Eating was once such a simple spiritual exercise, a covenant between earth and man and a nourishment of body and mind. Multinational food producing machines have turned it into a demonic act. Where eating a hamburger encompasses such evils as destroying tropical forests, enslaving and torturing animals, poisoning our surroundings, hardening our hearts and closing our minds. By making us accomplices in their acts, corporations insure themselves against recrimination. We are all responsible, so we refuse to acknowledge what we are doing. The most suppressed and subversive knowledge about the inner workings of our society involves breaking the spirits of fellow earth beings. Economic rationales excuse the enslavement of Africans, damming of rivers, killing of natives and the elimination of free living creatures. Laboratory animals are agonized by the scientific inquisition, tortured through dissection and made to confess that life is an object.

The least talked about horror is in the raising of animals for food. The conditions that factory farmed animals live in is so atrocious, that it makes the cruelty of the old family farm seem like a paradise. We turn our eyes away, still the shockwaves of agony felt by animals living lives of constant torment and despair shatters us from deep within. Crowded, debeaked, dehorned, prodded, belted, gutted, injected, intubated, mutated, diseased, distressed, stuffed, packaged and roasted and eaten. Who acknowledges the misery that these individuals must endure every day of their lives? We've condemned them to hell for the crime of being born. These shattered souls haunt the shadows, while they're tortured bodies reincarnate as our own flesh. Death is welcomed by those that suffer endlessly. Therefore factory farms accomplish something much worse then death, the renunciation of life. The death camps of Nazi Germany used the same methods of mass produced desouling. The faces of the survivors display the end product of this industry – emptiness.

From where to here she destined to remember the place. Voices hidden in burbles, eyes looking out, looking in from all knowing tree nodes. Wild irises gracing a forested brook, born in this world an instant of bright shadow. Roots down to the very center, sex spreading open to the suns blinding joy.

In her winter sleep awoken to a terrible elsewhere. Here ruled by a secret mimic a society of those skeptical of existence. An inverted habitat. See a place, see a race, demented apes enjailing the world. Night terror smart bomb, freak shrouded mechano-puppets. Poison stream leached into blood. A myriad of drowning clots trying to grasp at the sea, here where beaty loses. She a billion angels solidified and demoted in a place that insists on extinction.

A two-fold process of domestication and annihilation is practiced by the earth-razing machine. Buffalo, wolves, pronghorn and Indians are gunned down to clear the way for the domesticates. Cows, dogs, sheep and citizens are put to work towards greater conquest.

A strange relationship binds killer to victim, master to slave, first world to third. Abstract goals of endless profit and power obsess those that claim to be on top. Only to find out that the flesh that they slash and sell comes from their own core, ripped from their hearts and stolen from their children. For every killer that rises to power within our minds, a victim is thrown into the shadow awaiting retribution. Inside the molester hides the abused, inside the sadistic cop is the masochistic delinquent and within the CEO robber baron cry the starving multitudes. Only by forgetting that the inner reflects the outer, that we are all one being can we inflict such cruelties. An instant karma manifests itself when we are unjust to others, as the damage resonates inside us. What is ultimately best for the individual is also best for the whole.

It is a bizarre situation we have been thrown into. The world is a life-form that must feed on itself to survive, yet not destroy itself in the process. We must kill in order to eat, yet acknowledge the soul of that being as our own. The delicate relationship of predator and prey cannot survive as warfare, hatred or control, but only of partnership. Beings must give as much as they take for balance to occur, all else is theft. There is no top of the food chain, only a cycle in which we travel through all points.

The grand inquisitor arrived at the weathered entrance of the hermit.
"I've been expecting you", said the lone man.
"Whether a masochist or martyr, you are condemned to be broken" said the visitor.
"I am neither, for the pain you inflict on me, is that which you refuse to feel."

Placing the hermits hand in a specialized vice' the inquisitor twisted a rusty auger through the palm.

A world away the dreamer awakens, vicious red streaming the ends of his fingers.

Our health, both bodily and mentally is being compromised. The attack on our planet is a battle against our own body and mind. Disease, being the manifestation of imbalance follows our civilization as it devastates our world. We are taught that disease is an attack by an outside force on the body. Pasteur popularized the theory that microorganisms are responsible for illness. This theory is no longer as credible in this age where many diseases seem to be caused by the body attacking itself. Auto-immune disorders such as cancer, lupus, migraines, allergies, chronic fatigue, fibro-myalgia, multiple sclerosis, arthritis and hundreds of others have stumped the medical establishment. In one aspect, the new wave of disease is the logical outcome of releasing thousands of types of synthetic chemicals into the air, water and soil. These substances, unprecedented in our evolutionary history, enter into the balanced ecosystem of our body and overwhelm our ability to destroy or remove them. Our intelligent immune system becomes deranged as it struggles against a huge arsenal of alien poisons. Never having experienced chemicals such as PCB or DDT in billions of years, how can a liver manage to detoxify, or a kidney filter them from our blood.

To add further insult, many synthetics unlike natural chemicals don't degrade. Their structures refuse break down, so that they continue their damage long after they've been released. They concentrate in the bodies of predators. These substances also commonly mimic natural ones, so that the body unwittingly incorporates them into its structure, often leading to genetic damage and cancer.

From a larger perspective, the body waging war against itself is a microcosm of the dissociation of earth. As above, so below. It is the ruin of our world, internalized. Women, being closer to Earth suffer more often from auto-immune disorders as attacks on our planet are more directly felt by them. Cancer rates have gone from 1 in 27 in 1900 to close to 1 in 2 today. The rise of the chemical industry following WWII, has made cancer a common disease, soon to be major killer of Americans. Each year thousands of new synthetics are introduced into our bodies, a cruel experiment by greedy profiteers on earth inhabitants, who have little say as to what we're exposed to. Unfortunately, there is great profit to be made in disease, as people will pay anything to feel well again. Sickness is a goldmine. People weakened by these chemicals are also easier to control, as few have the energy left to rise up. The extent in which our minds are controlled is revealed by the fact that we are being poisoned and are not defending ourselves against it.

The idea of what health is has also changed. At one time it was believed to be the normal state of the body. Now people act as if it is something that needs to be achieved, engineered or consumed.

The enormous use of pesticides is keeping the earth and body in permanent, chronic imbalance. It is outright chemical warfare against our world. Insecticides, fungicides, herbicides and antibiotics, we are told treat the disease of creatures that threaten to overtake us. Technology emerges as the great hero, saving us from these plagues. Many are now beginning to realize that the infestation is not the disease, but a symptom of the underlying imbalance. Modern, monoculture farming is so at odds with the diversity of natural systems, it is disease waiting to break out. Thousands of genetically similar plants, crowded in one place, and planted in straight lines is an offense to the diversity and individuality of wild earth and an easy buffet for bugs. Pesticides temporarily keep the insects down, only to eventually make them stronger and more resistant. Furthermore all of the insects' predators are poisoned, leaving the target bug beyond natural control. In the suburbs of the northeast, pesticide use has resulted in epidemics of ticks, evolving in the Chem-lawn created vacuum. The ticks have a hard shell and are not easily killed. The birds and spiders that once prey on them have been wiped out, leaving the ticks to breed out of control. Lyme's disease and the auto-immune syndrome it provokes is the result.

Insects evolve quickly and become resistant to pesticides at a faster rate then new ones can be developed. Crop losses, due to insects have actually increased since the 1940's. Considering that the insects ate less crops prior to the use of synthetic pesticides, the resulting poisoning of our world seems ever more senseless, but ever so profitable. The massive aerial spraying of Vietnam and now Columbia point to an even more sinister use of pesticides, the extermination of native people and their way of life. Much of our chemical industry evolved out of Nazi Germany, particularly their chemical weapons program. Scores of mad scientists were brought to the U.S. to continue their work. Organophosphate pesticides originated as a nazi weapon and was used in death camps. One of those called Malathion was sprayed all over New York City against mosquitoes, and in Tampa against fruit flies. It kills by removing the sheath that covers nerves, it is the possible cause of multiple sclerosis when a fetus is exposed.

A large portion of our planets' surface is continually covered by toxic potions, We are taught to spray them all around our houses and lawns and they soak into our food and clothes. When combined with predator eradication programs, bio-engineering, nuclear testing and genocidal warfare, pesticide use is one more weapon in the act of murder-suicide we are inflicting on our world.

The use of anti-biotics is analogous to that of pesticides, except bacteria are the target. The outcome is much the same. Resistant strains evolve quickly after an antibiotics introduction. This means big money for pharmaceutical companies and bad news for the body. Antibiotics are no miracle drug, for individuals they provide a great tool for treating infections, but for mankind they just postpone the outbreak of even more powerful diseases. They mortgage the future for a quick fix today.

Bacteria, far from being an enemy we should be at war with, are the foundation of all multicellullar beings on earth. We have evolved as giant mega-colonies of bacteria. Within our cells they have evolved to become organelles. The diverse flora of bacteria far out number cells and are critical for our survival. Bacteria have the ability to digest any food, and a large variety that live in our intestines help to extract and create nutrients. They transfer genes amongst species and form communities that share information and nutrients. Antibiotics reduce this diversity and cause instabilities in the body. The word antibiotic means against life. They are used to treat every ailment and are put into the food of factory farmed animals. The super-strains of bacteria that have resulted have made us addicted to their use, as our bodies can no longer fight the monsters that our technology has mutated.

Every surface is a world trying to achieve balance. The use of anti-bacterial soaps, mouthwashes, cleansers and disinfectants ultimately achieve the opposite of their stated goals. By destroying diversity, they make way for disease. The vacuum left by sterility is soon filled by the most resistant of infectious agents.

Beings targeted for termination are labeled as pests, germs or weeds. Their worthiness of existence is denied and then a war is initiated to get rid of them. Much of technological innovation is created as part of this war. Networks of control are assembled, chemical agents prepared and the Frontline, of Combat is prepared to Raid.

An assumption is made that the diseases that plague man have always been with us. Research has shown that most diseases came to us as a result of domesticating animals. Keeping them in unnatural conditions have made the animals we control into incubators of disease. From cattle originated smallpox, measles, tuberculosis and diphtheria. Influenza came from pigs, the cold from horses and leprosy from water buffalos. Our own bodies became breeding grounds of harmful microbes as we moved into the crowded cities. From the rise of cities came cholera, dysentery, typhus and plague. Brought to the relative harmony of the Americas, these embodiments of imbalance swept a wave of death across these continents. As their cultures and spirits were destroyed by missionaries and pillagers the Native Americans were also attacked from within. It was an attack on all levels. Up to 90% of the inhabitants of the new world were killed by beings spawned within the bodies of the European invaders.

Improvements in diet, sanitation, and medicine have checked the spread of communicable diseases in many cities. New plagues of heart disease, asthma, and cancer have risen due to poor diet, stress and pollution. Some new diseases such as AIDS or Ebola might have originated in a laboratory along with hundreds of others as bio-weapons or as race specific eugenic instruments.

Due to pollution and other stresses many species of trees are dying off. In their weakened state their immunity breaks down and they succumb to infection by fungi and insects. An introduced fungus has killed all of the adult chestnut trees in the Northeast. These trees which use to feed the whole forest were the most common tree of many areas. They have been reduced to sprouting stumps. In a similar way the elm trees that once graced the streets of many villages have disappeared. Acid rain and other pollutants have weakened hemlock, spruce and fir trees and have left them open to attack by insects. Worldwide trees are suffering due to forest fragmentation, global warming and pollution.

The story of our civilization is one of going one step forward and two steps back. A time of rich evolutionary innovation and linear knowledge, combined with a gradual loss of wisdom and balance. What fails us most is the myth of Progress, also known as the Big Lie. Compromising less then 5% of human history and only describing the lives of privileged classes and populations, the myth of progress powers the machine of evolutionary disintegration. It has moved us away from the sacred stream of life and love and toward the disconnected backwater of nihilism.

Huge advances in the material wealth and comfort of a select elite have come at a painful price for the impoverished masses, whose lives are characterized by poor health and wage slavery. Obscene material progress for a few comes at the cost of starvation for many.

A decline in stature, health and life expectancy occurred with the advent of agriculture. This and the loss of egalitarian social structures were the first few steps toward the Fall. The need to work our whole lives away was imposed on us by the created hierarchies who directed a prolonged scarcity of all that once was free. Now we no longer know what we are working for or working on, we just keep working.

In the back of our minds we remember a Golden Age. A time when warfare was rare, diseases were few and we lived at peace with the land. Vestiges of this time still exist where wilderness prevails. There we can see beings that are vital and strong, living in a state that we strive for but rarely achieve. Grace of movement mixed with periods of serenity and noble patience. Even the violence of eating and the aggression rituals surrounding courtship are filled with beauty. Death and decay are admitted into the unity, as rebirth is assured in the motions of the spiraling helix. In the mindscape of wilderness, each being expresses itself according to an inexpressible order, where inner and outer rhythms synchronize. An ineffable state where self- interest and community meet and control is unnecessary. To find this place there is no need to travel in time or space, as it is also a place within. We arrive there when our thoughts are without words and our soul is set free to drift.

Traveling back to ancient Mesopotamia, one can see a microcosm of the cycle of destruction that we follow. First the land is cleared of all its native inhabitants, irrigation canals are built and cities rise up. Cruel leaders, full of arrogance and greed push the land and people until they break. Finally, the dissolution of everything in a frenzy of warfare and environmental collapse. When the land becomes saturated with salt from irrigation, the forests reduced to charred stumps and the cities crumble away due to war and erosion, the disease simply metastasizes to fresh areas of earth, yet uncontaminated.

A similar story occurred in the Indus valley 4000 years ago. The forests were burned away to dry bricks for the cities and the exposed soils were washed away. Waves of conquest sealed the fate of this land.

Spread by the march of empire, devastation followed in civilizations wake. All around the Mediterranean the fertile land was sucked dry. Many who visit the area now might assume that that the rocky infertile land of olive trees and scrub bushes was always there. It's hard to imagine the lush forests of evergreens, oaks and beeches or the diverse wildlife that once lived there. Following the clearing of the forests, overgrazing by cattle, goats and sheep assured that little would grow back and the soil would wash away. With the disappearance of cloud producing forests, rainfall diminished. Deserts expanded into the once fertile lands of North Africa, Israel and Turkey. Plato wrote in Critieas, " What remains compared with what existed is like the skeleton of a sick man, all the fat and soft earth having wasted away… but the soil it had was deep and therein received the water, storing it up in the retentive loamy soil." This was written before the destruction caused by the Roman Empire began. After Italy was laid bare, other provinces fell to the plow and ax. Libya was turned into a vast desert. The great cedars of Lebanon were reduced to a fraction of their grandeur. Only 10% of the forests that once stretched from Morocco to Afganistan still remain.

The same terrible sequence occurred in China, Japan, Ethiopia, ancient Mexico and coastal Peru. Every type of living system was attacked. Marshes were drained, rivers dammed silted and polluted, seas filtered of life and prairies carved up. Once upon a time places like Florida and California were true edens, not the parodies of paradise that they've become.

Worldwide, the stories of destruction follow a common narrative as each civilization follows the pathways dug by their forerunners. Only the costumes change and the technologies becomes more efficiently brutal.

New types of destruction began in our times. Attacks on our atmosphere, moved the war into the sky, as millions of smokestacks pump their toxins into the air. Holes in the ionosphere allow ultraviolet waves to cancer our skin, sulfuric acid tinged rain sicken the remaining trees, carbon ladened air give fever dreams of melted icecaps and rising floods. The U.S. Empire mimics Rome, pulping earths' beautiful life into its hungry vacuum of a culture. There is major difference in our story, WE HAVE NOWHERE LEFT TO RUN TO. We have no new frontiers to mar, the other planets are to distant and we are running out of time. The messiah we await arises from within. Only by saving our larger self may we be redeemed. The system we all obey robs us of meaning, knocks us off balance, destroys beauty and seeks to eliminate our future. Yet we still slave away, kneel in front of its glowing altars and domesticate our children towards submission. At once, we long to be free, yet we spend our lives placing chains. We want to do what's right but we're trapped in cycles that make us destroy. We can't reform the system with laws and boundaries, because the very existence of such things are part of the problem. If the system in which our lives revolve around is faulty and our beliefs are flawed, reform only serves to delay the immanent collapse.

The collapse of the outer order is a societal nervous breakdown. The more rigid the controls that hold someone together the greater the explosion. It is deeper truth colliding with a narrowed view of reality. It is the emergence of chaos into clockwork regularity, and paradox into indoctrinated thought. It is the movement of sickness out from the shadows and into conscious mind and thus it is the realization of our worst fears.

The initial symptoms of breakdown appear at first to be a dissociation of mind into separate pieces. On closer examination, it is revealed that these parts were lopped off at the time of repression and imprisoned in the darkness. Suppressed instincts, feelings and pathways warp hideously in their restrained state, they are the criminals created with the advent of laws. Their unleashing on the world or their burst into consciousness is associated with anarchy. This is only the first stage. It is the first step toward true healing. For most people and cultures that have experienced this breakdown, an overwhelming fear sends them running to any order that can shove the tortured souls back to their cells. This fear may cause a reactionary attempt to rebuild the old system. It may prod people towards a new order, set up by a charismatic leader, promising salvation and distributing a new doctrine of control.

Unleashed along with the demons are the angels. The joy and magic of being alive flower into being, as what once seemed like mundane existence turns to pure wonder. Sacred truths that hide beneath the seemingly solid veneer of matter are revealed.

The intricate network of connections that synchronize all aspects of existence become visible. Unity becomes known, not just as a concept, but felt in every cell of being. Then one can know what it feels like to truly be alive.

The idea that we must advance to a higher state of being is illusory, for deep inside we are already there. Consciously striving to create a utopia, to engineer a world is an empty goal pushed by foolish leaders. As we try to take control of our lives, we centralize power, splitting our ego from our Self. We assume the world cannot function without a ruling Lord, yet our bodies function without our control and a forest grows without a president. We're taught that it is human nature to want to seize power and rule over others. Reinforced by our history books and stories on the nightly news, we can hardly dream of a world at peace. When our own leaders plot attacks against us, when war, illness and starvation break us down, hope is hard to come by. It appears that we've been born into a universe ruled by evil, where no good deed goes unpunished and the righteous get crucified. We need to see that this trend is a trick of time. We are experiencing that part of the story where the antagonist has the upper hand, the heroes' life hangs perilously over a burning pit. Without perspective, lacking wholeness in our lives we feel doomed. As humanity enters into it's darkest night, we can find hope nestled in our center. It is the stillness of deep water as a storm batters the surface. Watch as a forest seedling cleaves out of devastated ground and know that the greatest path is toward reconciliation and harmony.

Made in the USA
Monee, IL
08 November 2026

67944874R00080